# PLANNING PRIMARY
# GEOGRAPHY

Other titles in the **Key Strategies** series:

# PLANNING PRIMARY

# GEOGRAPHY

## FOR THE REVISED NATIONAL CURRICULUM

## Key Stages
## 1 & 2

### Maureen Weldon
### Roy Richardson

**JOHN MURRAY**

Cover Photograph: Photoair Ltd
Line drawings (pages 74 and 79): Art Construction

© Maureen Weldon, Roy Richardson 1995

First published in 1995
by John Murray (Publishers) Ltd
50 Albemarle Street, London W1X 4BD

Layout by Christie Archer
Typeset by Anneset, Weston-super-Mare
Printed and bound in Great Britain by St Edmundsbury Press, Bury St Edmunds

A catalogue entry for this title can be obtained from the British Library

ISBN 0 7195 7062 X

# Contents

# CONTENTS

# What is good geography teaching?

## Setting standards

It should be every school's aim to provide children with the opportunity to develop their skills, knowledge and understanding of geography. We know that children learn best when they are interested and motivated to know more about what they are studying. The main responsibility of any teacher, therefore, is to develop this enthusiasm and interest in the subject. As children have a natural interest in people and a curiosity about places, developing and extending their interest in geography ought to be easy! The challenge for all schools is to develop within children a greater understanding of the fascinating world in which they live, and a love of geography that will extend throughout their adult life.

But how is this love and enjoyment of geography to be developed in young children?

In any primary school it is the headteacher who most influences the ethos and curriculum of the school. In a school where good practice in geography exists, the headteacher will have ensured that:

■ School policy clearly states how geography is to be taught throughout the school.

■ A Programme for Geography exists, which sets out how the statutory orders for geography will be covered throughout Key Stages 1 and 2.

■ A Scheme of Work informs teachers' own planning.

■ Staff are well-trained and knowledgeable.

■ There are adequate resources to promote an enquiry approach to geography.

■ An assessment procedure informs teachers' own planning.

**A journey round the school**

Where a school has provided well for the teaching and learning of these elements, a visitor to the school might expect to see a wide range of activities being undertaken throughout the school year. On such a visit you might see. . .

### RECEPTION

In the Reception class the teacher has taken the children on a journey around the school and identified certain locations (a selection of classrooms and the secretary's office). The teacher is now sitting with the whole class talking about their school and how they could go about finding their way to the locations already identified. Some of the children are giving simple directions using the terms 'next to', 'near' and 'along'. Some of the children are showing the route they would take from their classroom to another by tracing their journey on the large-scale map that the teacher has drawn. There are photographs of different parts of the school that are familiar to the children. Some children can identify the photographs and locate them on the class map.

### YEARS 1 & 2

The Year 1 and 2 classes are preparing to go on a walk in the area adjacent to their school. They have talked about what they know about the area around their school and the information has been made into a classroom display by the teacher. The display includes a large map of the local area that the teacher has drawn. The map is being used by some children to describe the direction from one locality to the next. The places the children have identified have been labelled by the teacher. The older children have copies of the map to follow on their walk. On the walk, the teachers intend to focus the children's attention on the way in which the land and buildings in the area are being used, so the children will be collecting evidence with this in mind. During the walk, photographs of particular features will also be collected to form a classroom display. There is a computer in use, with two children developing an understanding of directional terms using *The Playground* by Topologika.

### YEARS 3 & 4

In Years 3 and 4 the children are studying Kaptalamwa, a village in Kenya, using a photopack purchased from the Geographical Association. After looking closely at the photographs provided, some children are compiling a list of the similarities and differences between Kaptalamwa and where they live. Several children make reference to the features they photographed

Kaptalamwa

Kaptalamwa is a small village in Kenya there ain't many people compared to us. The schools have no windows and they are made out of wood. There are gaps in between the pieces of wood. The houses are made out of clay and are usually painted. The houses can last for 50 years. At the top of the houses some people store crops. They get wood from trees to make a fire and a fence. They have to go to streams to get water. Children can't go to secondry school before they're clever enough.

around the school when they were in the infants'. Another group are using a globe and an atlas to help them to plan a journey from their school to Kaptalamwa – focusing upon direction and countries along the route. Two children are using an art programme on the computer to draw a map of Kaptalamwa village, drawing upon the information in the pack.

### YEAR 5

Year 5 are not in school this week. They are on a school visit to Boggle Hole in Yorkshire. The classroom door is open and you can see that a great deal of work has been undertaken to prepare for this annual visit. Children have been using Ordnance Survey maps (1:25 000) to identify the main human and physical features around Boggle Hole. Using a road atlas, another group have mapped out the route they think the coach will take from their school to Boggle Hole. This group has also estimated the distance and time it will take for the journey and where they might need to stop for lunch! Whilst away, the children will collect a great deal of geographical information by recording, making field sketches, photographing physical and human features, talking to local people and collecting information leaflets, which they will use to continue their study back at school. The headteacher has received several postcards from the children saying how nice it is not having to do any work!

## YEAR 6

The Year 6 class are undertaking a study of the weather. The children had undertaken a weather study in Year 2 and are used to making recordings over a long period of time. Each week, a group of children has been responsible for making measurements of temperature, rainfall, wind speed and direction and cloud cover at three different locations around the school. The records have been placed into the class Weather Book. Children enjoy looking back at the records to identify extremes in weather conditions. The class have been looking at the relationship between rainfall and cloud cover, wind direction and temperature, and the influence of different locations on their weather data. Several children explain to you that their records show that the temperature is usually lower when the wind blows from the north or the east. There is a school weather station attached to the class computer, which automatically records weather data each day. The children have discussed the advantages and disadvantages of the different recording systems.

This school expects that, by the end of Key Stage 2 the children:

■ will be asking questions about places, and will be eager to find out more.

■ will know more about their local area, and will be able to relate certain aspects to other parts of the UK and the world. They will have begun to appreciate how people can shape the world for both good and bad.

■ will have begun to show tolerance by their readiness to accept differences between cultures.

■ will have developed feelings of responsibility towards others and the environment.

■ will have developed their geographical skills, using maps and globes as a regular part of their work, and be confident in using geographical vocabulary.

■ will, most importantly, be enthusiastic about geography and will have been given opportunities to be geographers.

## Figure 1: Elements of primary geography

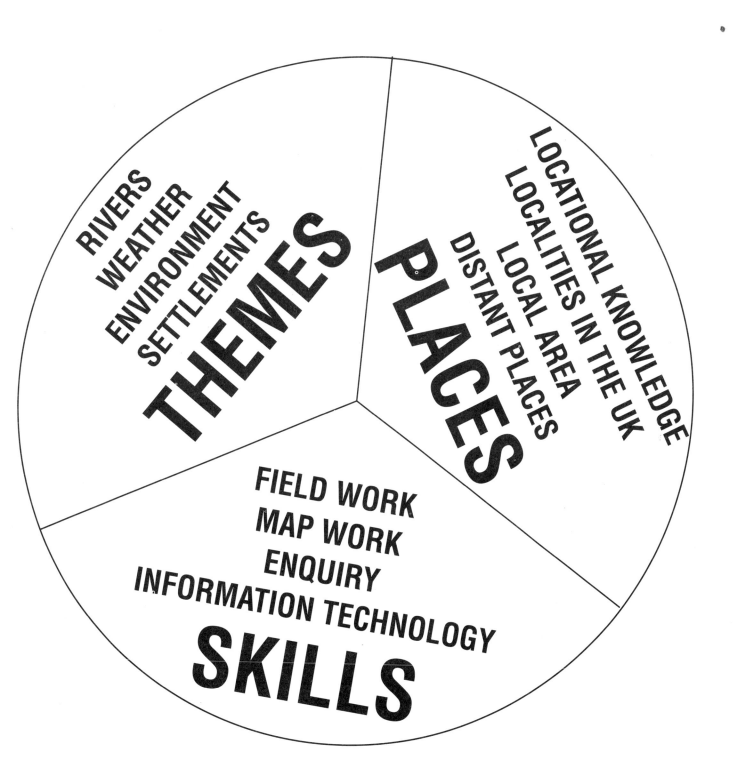

THEMES

RIVERS
WEATHER
ENVIRONMENT
SETTLEMENTS

PLACES

LOCATIONAL KNOWLEDGE
LOCALITIES IN THE UK
LOCAL AREA
DISTANT PLACES

SKILLS

FIELD WORK
MAP WORK
ENQUIRY
INFORMATION TECHNOLOGY

# What are the statutory requirements?

## Statutory requirements for schools when planning geography

Schools should plan to implement the new statutory requirements for geography from August 1st 1995. This section gives an overview of the changes required by the new Orders. It also sets down the new and old Orders in chart form to enable you to see, at a glance, exactly where the changes have been made. As the majority of schools will already have plans in place for geography, it is hoped that these charts will help you to review these existing plans, rather than to draw up completely new plans.

Overall, the new Orders are to be welcomed by primary teachers. They are presented in a clearer format, and the new wording makes them easier to understand. The challenge for schools is in ensuring that all teachers have a clear understanding of the Statutory Orders, and that the Orders are then customised to meet the needs of the children.

Figure 2 presents an overview of the key changes in the new Order.

Figures 3 and 4 summarise what the statutory requirements now are.

These summaries of the old and new Orders should prove useful to all teachers, but particularly to geography co-ordinators. The charts may be photocopied and used in staff meetings. They can also act as useful references when teachers are involved in whole school planning.

Note that the charts do not use the exact wording of the Statutory Orders, but provide a summary for reference. However, no changes in meaning or interpretation have been made.

# Figure 2: Overview of the key changes in the new Orders

■ A single attainment target – called Geography – replaces the five attainment targets. This single attainment target encompasses elements of skills, places and themes (physical, human and environmental) from the original Order.

■ Programmes of Study set out the knowledge, understanding and skills to be taught with distinctive requirements for each Key Stage.

■ Overall there has been a substantial reduction in statutory content at Key Stages 1 and 2.

■ A significant change has been the replacement of the statements of attainment for geography with level descriptions.

■ The new Orders have been restructured at each Key Stage to aid clarification and improve progression within and across Key Stages.

■ Progression in individual components of geography can be traced through the level descriptions.

■ The locational knowledge children need has been integrated into the general requirements and the skills section of each Key Stage.

■ Maps have been redrawn using clear criteria for including places. A different map projection – Ekert IV – has been used for Map C.

**Main changes at Key Stage 1**

■ Place studies have been cut from three to two. These are the locality of the school and a contrasting locality, either in the United Kingdom or overseas.

■ The number of themes (the current 'strands') have been reduced from nine to one, which is about environmental quality.

**Main changes at Key Stage 2**

■ Place studies have been cut from five to three: the locality of the school (covering an area larger than in Key Stage 1), another locality in the United Kingdom and a third in Africa, Asia (excluding Japan), South America or Central America (including the Caribbean).

■ The number of themes has been reduced from eleven to four: Rivers, Weather, Settlement and Environmental Change.

# Figure 3: The statutory requirements for Key Stage 1

## The new Orders – Key Stage 1

1. *Pupils should be given opportunities to:*
   a. investigate the human and physical features of their surroundings;
   b. undertake studies that focus on geographical questions, based on direct experience, practical activities and field work. Studies should involve development of skills, and knowledge and understanding of places and themes;
   c. become aware that the world extends beyond their own locality, in and outside the United Kingdom, and that the places they study exist within this broader geographical context.

### GEOGRAPHICAL SKILLS

2. In investigating places and a theme, pupils should be given opportunities to observe, question and record, and communicate ideas and information.

3. *Pupils should be taught to:*
   a. use geographical terms in exploring their surroundings;
   b. undertake field work activities in the locality of the school;
   c. follow directions, including the terms up, down, on, under, behind, in front of, near, far, left, right, north, south, east and west;
   d. make maps and plans of real and imaginary places, using pictures and symbols;
   e. use globes, maps and plans at a variety of scales; identifying major geographical features, locating and naming on a map the constituent countries of the United Kingdom, marking on a map approximately where they live, and following a route;
   f. use secondary sources to obtain geographical information.

### PLACES

4. Pupils should study the locality of the school (school buildings, grounds and surrounding area within easy access) and a contrasting locality, either in the United Kingdom or overseas (similar size).

5. *In these studies, pupils should be taught:*
   a. about the main physical and human features;
   b. how localities may be similar and how they may differ;
   c. about the effects of weather on people and their surroundings;
   d. how land and buildings are used.

### THEMATIC STUDY

6. The quality of the environment in any locality, either in the United Kingdom or overseas, should be investigated.

   *In this study, pupils should be taught:*
   a. to express views on the attractive and unattractive features of the environment concerned;
   b. how that environment is changing;
   c. how the quality of that environment can be sustained and improved.

# What has gone from Key Stage 1

## GEOGRAPHICAL SKILLS

*From the range of skills specified:*
- Observe, describe and record the weather over a short period ———— *Deleted*
- Elements of study for pupils working towards level 3 ————

## PLACES

*From the study of places:*
- A contrasting locality in the United Kingdom
- A locality beyond the United Kingdom ————

*These are given as choices in the new Order*

Elements of study related to these localities, including elements for pupils working towards level 3 ———— *Deleted*

A framework of specified locational knowledge, including points of reference on Maps A and C ———— *Deleted*

## THEMES

*From the study of themes:*
- Rivers, seas and oceans
- Landforms, soil and rocks
- Settlements, communications, economic activities
- Natural resources

*Deleted*

# Figure 4: The statutory requirements for Key Stage 2

## The new Orders – Key Stage 2

1. *Pupils should be given opportunities to:*
   a. investigate places and themes across a widening range of scales;
   b. undertake studies that focus on geographical questions; involving field work and classroom activities; development of skills, knowledge and understanding about places and themes;
   c. develop the ability to recognise patterns, explain patterns;
   d. become aware of how places fit into wider geographical context.

### GEOGRAPHICAL SKILLS

2. *In investigating places and themes, pupils should be given opportunities to:*
   a. observe and ask questions about geographical features and issues;
   b. collect and record evidence to answer questions;
   c. analyse evidence, draw conclusions and communicate findings.

3. *Pupils should be taught to:*
   a. use appropriate geographical vocabulary;
   b. undertake field work, the use of instruments for measurements and techniques;
   c. make plans and maps at a variety of scales, using symbols and keys;
   d. use and interpret globes, maps and plans at variety of scales. Includes co-ordinates, four-figure grid references, measuring distance and direction, following routes, contents and index of an atlas, identifying points of reference on Maps A, B and C;
   e. use secondary sources of evidence – including pictures and photographs, including aerial;
   f. use information technology for information and evidence.

### PLACES

4. Pupils should study the locality of the school (area larger than immediate vicinity of school); a contrasting locality in the United Kingdom and a locality in a country in South or Central America, Africa or Asia (excluding Japan).

5. *In these studies, pupils should be taught:*
   a. about main physical and human features, and environmental issues;
   b. how the localities may be similar and how they may differ;
   c. how features of localities influence the nature and location of human activities;
   d. about recent or proposed changes in the localities;
   e. how the localities have a broader geographical context, and link with other places.

### THEMATIC STUDIES

6. Studies should be set within context of actual places; some should use topical examples.
   Taken together: range of scales (local to national), range of contexts (UK, EU and world).

**7.** Rivers

*In studying rivers and their effects on the landscape, pupils should be taught:*

a. rivers have sources, channels, tributaries and mouths, receive water from a wide area; most eventually flow into a lake or a sea;

b. rivers erode, transport and deposit materials, producing particular landscape features.

**8.** Weather

*In studying how weather varies between places and over time, pupils should be taught:*

a. how site conditions can influence the weather;

b. about seasonal weather patterns;

c. about weather conditions in different parts of the world.

**9.** Settlement

*In studying how settlements differ and change, pupils should be taught:*

a. that settlements vary in size and characteristics, and locations reflect types of economic activities;

b. how land in settlements is used in different ways;

c. about a particular issue arising from the way land is used.

**10.** Environmental change

*In investigating how environments change, pupils should be taught:*

a. how people affect the environment;

b. how and why people seek to manage and sustain their environment.

## What has gone from Key Stage 2

### GEOGRAPHICAL SKILLS

From the range of skills, specified elements of study for pupils working towards level 5

*Deleted*

### PLACES

*From the study of places:*
- Home region
- A locality in an EU country outside the United Kingdom

*Deleted*

Elements of study related to these localities, including elements for pupils working towards level 5 — *Deleted*

From the framework of specified locational knowledge, points of reference on Maps D, E, and F — *Deleted*

### THEMES
*From the study of themes:*
- Landforms
- Soil and rocks
- Population
- Communications
- Economic activities
- Natural resources

*Deleted*

# Whole school planning

Once teachers have a clear understanding of the new Orders for geography, plans should be set down showing clearly when and how each element is to be taught throughout Key Stages 1 and 2.

Sections 4 and 5 in this book deal specifically with mapping out geography throughout the six years of primary education, and an example Scheme of Work (see pages 35-67) is provided for schools to adapt.

The following pages set out some broad principles that need to be considered before beginning the lengthy process of whole school planning of geography. It is hoped that these pages will prove useful to geography co-ordinators. Each section of information is set out on a separate page so that they can be photocopied and used in staff meetings, if so desired.

Level descriptions have replaced the old statements of attainment. The level descriptions for geography are shown on page 16. Unlike the old SOAs, the level descriptions are not intended to be used as a planning tool, but are for summative assessment at the end of the Key Stage.

The level descriptions describe the types and range of performance which children working at a particular level should characteristically demonstrate. The great majority of children will be working at levels 1 to 3 by the end of Key Stage 1, and at levels 2 to 5 by the end of Key Stage 2.

See page 82 for a detailed discussion of how the level descriptions are to be used.

# Figure 5: Planning your geography curriculum – key points

**At Key Stage 1**

■ Children's learning should be based largely on direct experience, practical activities and field work in the locality of the school.

**At Key Stage 2**

■ Each of the five themes can be taught as part of the study of places, in combination with other themes, or alone. Whichever option your school chooses, thematic work should be set within the context of actual places, and some should use topical examples.

Thematic studies should involve work at local and national scales. A range of geographical contexts in different parts of the world, including the UK and the European Union, should be used.

**At both Key Stages 1 and 2**

Each geographical study should involve the development of skills, as well as knowledge and understanding about places and themes.

■ Planning should take account of the need to develop children's awareness of the world extending beyond their own locality and of the broader geographical context for their locality studies.

■ Children should be encouraged to develop their skills, knowledge and understanding through geographical questions.

# Figure 6A: Planning for progression (1)

■ Progression is the careful and deliberate sequencing of learning so that previous experience can be built on, and future learning prepared for.

■ It is necessary to understand the progression in geography in order to put together your programme and scheme of work for geography.

■ Previous guidance on progression in geography, provided by the National Curriculum Council, is helpful to consider when planning.

---

Progression in geography is about offering children a sequence of work which gradually increases the:

■ level of difficulty of practical and intellectual tasks;

■ range of scales studied;

■ breadth of studies;

■ depth of studies;

■ complexity of phenomena studied and tasks set;

■ understanding of generalised and abstract matters as opposed to the concrete and specific;

■ awareness of issues involving different attitudes and values.

---

■ Realistically, you may focus on perhaps two or three of these goals over a short series of lessons, building on most of them over a one or two year plan.

■ Progression is essentially a matter of medium and long term planning.

■ In order to plan a programme and scheme of work for geography successfully you need to know what teaching and learning has taken place before, and what will come after, so planning should take place across a Key Stage.

# Figure 6B: Planning for progression (2)

**GEOGRAPHICAL SKILLS:**

■ Geographical skills are on-going and taught throughout the Key Stage.

■ To plan for progression in geographical skills, teachers may need to provide a sequence of steps towards the acquisition of the skill.

■ Plan for the use which is made of these skills and the contexts within which they are applied: integrate the skills with places and thematic work being undertaken.

**THROUGHOUT KEY STAGE 1 CHILDREN WILL INCREASINGLY:**

■ Recognise and describe what places are like, using appropriate geographical vocabulary.

■ Offer their own views and judgements about what they observe.

■ Make comparisons between places and between geographical features.

**THROUGHOUT KEY STAGES 1 AND 2, CHILDREN WILL INCREASINGLY:**

■ Broaden and deepen their knowledge and understanding of places and themes.

■ Develop and use appropriate geographical skills.

**THROUGHOUT KEY STAGE 2 CHILDREN WILL INCREASINGLY:**

■ Recognise and describe what places are like with accuracy and coherence.

■ Offer explanations for the characteristics of places.

■ Identify physical and human processes and describe some of their effects.

■ Apply geographical ideas learnt in one context to other studies at the same scale.

■ Acquire information, from secondary sources as well as first-hand observation, to investigate aspects of local and more distant physical and human environments.

# The level descriptions

### LEVEL 1

Pupils recognise and make observations about physical and human features about places. They express their views on features of the environment in a locality that they find attractive or unattractive. They use resources provided and their own observations to respond to questions about places.

### LEVEL 2

Pupils describe physical and human features of places, recognising those features that give places their character. They show an awareness of places beyond their own locality. They express views on attractive and unattractive features of the environment of a locality. Pupils select information from resources provided. They use this information, and their own observations, to ask and respond to questions about places. They begin to use appropriate vocabulary.

### LEVEL 3

Pupils describe and make comparisons between the physical and human features of different localities. They offer explanations for the locations of some of those features. They show an awareness that different places may have both similar and different characteristics. They offer reasons for some of their observations and judgements about places. They use skills and sources of evidence to respond to a range of geographical questions.

### LEVEL 4

Pupils show their knowledge, understanding and skills in relation to studies of a range of places and themes, at more than one scale. They begin to describe geographical patterns and appreciate the importance of location in understanding places. They recognise physical and human processes. They begin to show an understanding of how these processes can change the features of places, and that these changes affect the lives and activities of people living there. They describe how people can both improve and damage the environment. Pupils draw on their knowledge and understanding to suggest suitable geographical questions for study. They use a range of geographical skills – drawn from the Key Stage 2 or Key Stage 3 programme of study – and evidence to investigate places and themes. They communicate their findings using appropriate vocabulary.

### LEVEL 5

Pupils show their knowledge, understanding and skills in relation to studies of a range of places and themes, at more than one scale. They describe, and begin to offer explanations for, geographical patterns and for a range of physical and human processes. They describe how these processes can lead to similarities and differences between places. Pupils describe ways in which places are linked through movements of goods and people. They offer explanations for ways in which human activities affect the environment, and recognise that people attempt to manage and improve environments. Pupils identify relevant geographical questions. Drawing on their knowledge and understanding, they select and use appropriate skills (from the Key Stage 2 or Key Stage 3 programme of study) and evidence to help them investigate places and themes. They reach plausible conclusions and present their findings both graphically and in writing.

# Writing your Geography Policy

## How will the policy support geography teaching in your school?

The writing and review of your Geography Policy will encourage professional debate and increase the staff's awareness of what is required in their teaching.

It will help to develop continuity and progression, and give guidance for maintaining and developing quality in teaching and in the work the children undertake.

It should give guidance to new staff who, having read it, should be clearer about the school's expectations of them.

It should also guide the headteacher, co-ordinator and any visitors in what to look for when evaluating the quality of the geography teaching taking place in the school.

### The curriculum policy for geography

The policy should:

■ Reflect the ideas and philosophy which are promoted in your school's whole school curriculum policy.

■ Reflect the ethos which you are attempting to create for the teaching of geography at your school.

■ Guide the class teachers when planning a range of geography activities for all children.

■ Help achieve consistency.

■ Be short, and written in a language everyone can understand –
i.e. parents and governors.

■ Reflect the work currently being undertaken in the classroom or what the whole staff are working towards.

■ Help the class teacher to plan to deliver the National Curriculum.

### Guidance for drawing up a Geography Policy

The staff need to establish an understanding of what constitutes good geography practice, and become aware of what the statutory orders are asking them to plan to cover. It is then possible for them to make policy decisions relating to how and when the children will be introduced to the different aspects of National Curriculum geography.

Once policy decisions have been made, it is easier for schools to identify how and when the various parts of the statutory orders will be taught.

## What makes a good policy?

A good policy would meet the following criteria:

■ It should be short and interesting so that people will want to read it.

■ Its authors have considered who its readership is, for example governors, teachers, parents, visitors to the school and inspectors.

■ It is free from unnecessary geography-specific terminology, especially if it is going to be distributed to non-teachers.

■ A teacher new to the school should be able to understand fully what is expected.

■ The policy shows clearly what you are setting out to achieve, and allows you to be able to identify or measure your achievements.

Remember that a good policy is a management tool which can assist headteachers in bringing about change in the quality of education within their school.

A good indicator of how well a policy has been prepared and written is to ask whether a visitor to the school could read the policy and, if everybody were teaching geography on that day, clearly observe aspects of the policy being addressed, being worked towards or set down in the school's long term plan.

## Questions to consider in drawing up a Geography Policy

Set out, on the following pages, are a number of headings which can be used to form the framework for a Geography Policy. Under each of the headings there is guidance on the types of issues that might be discussed at staff meetings. Teachers should be given the opportunity to discuss the points set down under each of the headings. They may also think of issues that are not included here, but are felt to be particularly relevant to your school.

At the end of this section is an example Geography Policy (see pages 22–26). Co-ordinators and headteachers have found that this example policy helps them to plan and guide the discussions at staff meetings. You may, however, choose to adapt the example policy to meet your own situation and requirements. This is particularly useful where a school does not feel it has the time available to hold a series of discussion meetings with staff.

### ADAPTING YOUR EXISTING POLICY

In many schools there may already be a policy for geography in place. If so, the publication of the new orders offers you an opportunity to review, as well as revise, your existing geography plan. The headings for discussion in this section are intended to reflect the fact that, in many cases, teachers will be reviewing an existing policy. However, they are equally relevant to any school drawing up an entirely new Geography Policy.

Before you begin, it will be useful to think about:

■ How was the policy drawn up? Were all staff involved? Did all the staff agree to its content?

■ What was the role of the co-ordinator in drawing up the policy?

■ What are its strengths and weaknesses?

## 1. BACKGROUND

How are staff to be made aware of the content of the policy? How does the school ensure that new staff and supply teachers are aware of the policy?

Where can copies of the policy be found?

Are parents given access to the policies? Is any other body/person allowed to request a copy ? What are the procedures for requesting a copy ?

Is there any documentation that you feel should be attached, as an 'Appendix', to the policy ? (For example, the school's Programme for Geography, resource lists, the school's policy on visits, etc.)

## 2. THE PHILOSOPHY OF GEOGRAPHY

What do the staff feel that geography is about? What are the main aspects of geography that the children in your school should have experienced and learnt about by the time they leave?

How do you feel geography should be taught within your school? Should emphasis be given to any particular aspect of geography?

What studies will be undertaken at Key Stage 1?

What studies will be undertaken at Key Stage 2?

Will you make reference to the school's Programme for Geography, so that details of what is to be covered at each Key Stage can be kept to a minimum in the policy?

## 3. GEOGRAPHY IN THE NATIONAL CURRICULUM

Will you write a short paragraph which explains the geographical skills, places and thematic study sections?

Will you explain what the statutory requirements are at each Key Stage?

Will you give more specific detail of anything you wish to emphasise at each Key Stage?

Has the school drawn up a Programme for Geography which sets down exactly when various aspects of the Statutory Orders are to be covered? This may be attached to the policy as an Appendix.

## 4. TEACHING STRATEGIES AND PLANNING

What guidance do teachers have to assist their planning for geography? Is there a Scheme of Work for geography? How will the Scheme of Work assist the teachers in their planning and teaching?

What approach will staff be required to take in planning the study of places?

Are there any specific links with other areas of the curriculum? Are these indicated in the Scheme of Work?

Are there any aspects that the school wishes to make central to the teaching of geography? Is emphasis given to first hand experience, practical activities and field work?

Is the investigative approach to geography emphasised?

Are there specific links with the local community that need to be emphasised? Is there a particular industry that plays a leading role within the community that should be involved within geography studies wherever possible? Have specific links been set down within the Scheme of Work?

Does the school take the children on a residential field trip each year? Are there any other trips that classes take each year that have a geography focus?

How are the geographical skills, including map work, to be developed?

Does the school have any computer software or hardware that is particularly relevant to the teaching of geography?

## 5. IN THE CLASSROOM

Will there be any specialist teaching for geography? Will specific aspects of geography be taught by somebody with a specialist knowledge?

Is there a range of teaching styles adopted that is matched to whatever is being taught? Is this identified within the Scheme of Work?

What types of practice would the school expect to see when geography is being taught?

What is meant by investigative geography in terms of how geography is taught in the classroom?

## 6. EQUAL OPPORTUNITIES AND SPECIAL NEEDS

How will the school ensure that all children are given the same opportunities within their geography work?

How will teachers plan to ensure that all children are having opportunities to work to their full potential – whether they are the least or most able?

How are staff expected to provide access to the geography curriculum for those children with special educational needs?

Does geography have any specific role in the breakdown of stereotyping within ours and other cultures?

Does the school's plans for geography develop a greater understanding of the customs and cultures of others?

## 7. ASSESSMENT AND RECORD KEEPING

How and when are teachers expected to assess children's progress in geography?

What other guidance exists in school on assessment?

How are the level descriptions to be used in undertaking assessments?

Does any guidance exist on interpreting the level descriptions? Does the school have a portfolio of children's work that offer guidance on making assessments in geography?

As the level descriptions are for end of Key Stage assessments, has the school made any provision to make assessments midway through Key Stage 2?

## 8. RESOURCES

What aspects of information technology will contribute to the teaching of geography? When, and within which studies, will each aspect of information technology be most relevant? Is this information set down in the Scheme of Work?

How are the geography resources to be organised within the school?

Who is responsible for the resources, the drawing up of the resource list and the ordering of new equipment? How are teachers to order geography equipment?

Are resources to be kept centrally as well as in each classroom? Are the arrangements different within each Key Stage?

Is there a system adopted for the borrowing and return of central resources?

How will children be encouraged to use the resources?

### 9. EARLY YEARS

How will the very young children in school be introduced to geography and the way of working? How will the school ensure that the work they undertake leads in to, and is built upon, at Key Stage 1? What are the links between the Reception plans and those at Key Stage 1?

### 10. SAFETY AND CARE

What school guidance exists on taking children out of school or on residential trips? Where can this information be found?

What local authority guidance exists on school trips? Are there any aspects of the guidance that need to be set down in the policy?

Are there any aspects of the school's Behaviour Policy that the staff feel are specific to children attending school trips?

Where is the school's Health and Safety Policy kept?

Do any procedures exist to ensure that all staff, new teachers and supply teachers are fully aware of the procedures to adopt when taking children on school trips or residential weekends?

### 11. REVIEW

How often will you as a school review your policy to update and refine it, as appropriate?

# Figure 7: An example Geography Policy

For the purpose of this publication we have written an example school Geography Policy based upon the questions outlined on pages 18-21.

This example policy will form the basis upon which we will map out the whole of the Statutory Orders for geography. It will also influence how we plan and teach geography and inform new teachers of what the school's expectations are with regard to the teaching of geography. The policy will assist teachers in drawing up the Scheme of Work (see Section 5) and will be clearly reflected throughout the scheme.

The ultimate test of the success of your policy is that it is reflected within the practice that is observed throughout your school.

## 1. BACKGROUND

1.1 Geography is a foundation subject within the National Curriculum. This policy outlines the purpose, nature and management of the geography which is taught and learned in our school.

1.2 The school policy for geography reflects the agreed views of all the staff of the school. It has been drawn up as a result of a series of whole staff meetings led by the geography co-ordinator. The policy has the full agreement of the governing body and staff of the school. The policy was agreed by the governing body at their meeting of (date).

1.3 All staff have read and agreed the Geography Policy, and are fully aware of their role in its implementation. All staff have a copy of the policy in their school policy folders.

1.4 All new members of staff are provided with policy folders, and curriculum co-ordinators have the responsibility for explaining the teaching implications of their own policies.

1.5 Copies of all policies are kept in folders in the secretary's office and the head's room. Parents requesting to see a copy of the policy can do so by making their request to the headteacher.

The following are attached to this policy:

Appendix 1 – the school's Programme for Geography.

Appendix 2 – the school's geography resource list.

Appendix 3 – procedures for taking children on school trips.

## 2. THE PHILOSOPHY OF GEOGRAPHY

2.1 Geography is about the study of places, the human and physical processes which shape them, and the people who live in them. Geography helps us to understand the ways of life and cultures of people in other places.

2.2 Throughout the school, children will study their school locality. The area studied will expand as the children's understanding of geography increases. Comparisons will be made between the school locality and other localities within the United Kingdom. Children will also study the European Union and other parts of the world.

2.3 The study of our school locality forms an important part of the geography taught in our school, particularly at Key Stage 1. The school locality is defined within the school's Scheme of Work. The children's understanding and awareness of their school locality is developed through direct experience, practical activities and field work.

2.4 The teaching and learning of geography in our school should be both motivating and stimulating. Children should develop both knowledge of the subject and an enjoyment for undertaking further work in geography.

## 3. GEOGRAPHY IN THE NATIONAL CURRICULUM

3.1 In the National Curriculum, geography is set down under one attainment target called Geography.

3.2 At Key Stage 1, children are required to carry out three geographical investigations. Two investigations will focus on particular places – the locality of the school and a contrasting locality. The third focuses on a particular geographical theme. All investigations should involve the development of skills, as well as a knowledge and understanding of places and themes.

3.3 At Key Stage 2, children are required to carry out studies of three places. They are also required to study, in the context of actual places, four geographical themes. Elements of skills, places and themes should feature in all geographical studies, whether the main focus is a place or a theme.

3.4 All children should have an entitlement to access to the Programmes of Study, matched to their knowledge, understanding and previous experience.

3.5 Coverage of geography is set down in the school's Programme for Geography. The Programme identifies the skills and knowledge to be addressed within each study. The Programme ensures that all statutory requirements for geography are addressed.

3.6 The school's Programme for Geography is attached to this policy as Appendix 1.

## 4. TEACHING STRATEGIES AND PLANNING

4.1 It is important that the teacher identifies the most appropriate teaching strategy to suit the purpose of a particular learning situation. The Scheme of Work provides guidance on the most effective methods for teaching specific areas of study.

4.2 At Key Stage 1, much of the pupils' learning in geography will be based upon direct experience, practical activities and field work in the locality of the school.

At Key Stage 2, children will develop their skills, knowledge and understanding through geographical enquiries, across a widening range of scales, based on field work and classroom activities.

The locational knowledge on Maps A–C will be linked with place studies wherever possible.

### ■ Themes

There will be an emphasis on environmental geography at Key Stage 1 and a balance between physical, human and environmental geography at Key Stage 2.

### ■ Skills

All aspects of geographical skills, both general and specific, will be integrated within work on places and themes, as set down on the school's Programme for Geography. In planning, teachers will refer to the details of the geographical skills.

An investigative approach to geography will be taken throughout, with children actively participating in enquiry, field work, map work and the use of information technology when appropriate.

### ■ Map work

Children will be provided with opportunities to work with a wide range of maps, including historical and thematic maps at a variety of scales. Emphasis is given to developing the mapping skills of location, symbols, scale, perspective, style, drawing and map use. Children in Years 5 and 6 use 1:25 000 and 1:50 000 Ordnance Survey maps in their local area study.

4.3 A residential field trip is undertaken each year during the Summer Term with the Year 5 class. The field trip will focus primarily upon developing geographical knowledge and skills. The aspects of geography that will be focused upon during the field trip are set down within the school's Scheme of Work.

4.4 The local community and industry are involved within geography studies whenever possible.

## 5. IN THE CLASSROOM

5.1 Children are taught in their normal class group for geography

5.2 All teachers are responsible for the teaching of geography.

5.3 Teachers should look for opportunities to praise co-operation and safe, considerate behaviour.

5.4 Children are encouraged to work as individuals, in pairs, in groups and also as a whole class when appropriate.

## 6. EQUAL OPPORTUNITIES AND SPECIAL NEEDS

6.1 Activities both within and outside the classroom are planned in a way that encourages full and active participation by all children, irrespective of ability.

6.2 In our studies of localities equal emphasis will be given to the roles of both men and women at all levels of society. In distant locality studies our focus will be on the lives of real people and families to avoid stereotyping.

Every effort will be made to ensure that activities are equally interesting to both boys and girls.

6.3 Places studied should present opportunities for the children to gain an understanding of environments which contrast with their own.

## 7. ASSESSMENT AND RECORD KEEPING

7.1 Teachers will make brief notes on children's progress in geography at the end of each year. The notes should be kept to a minimum and yet provide enough information to inform the next teacher of the progress made, and to be of use in preparing the annual report to parents.

All assessments and records should comply fully with the school's Assessment Policy. Copies of the Assessment Policy are placed in teachers' own policy folders. Copies are available from the school secretary.

7.2 All teachers will be responsible for ensuring that assessments are made at the end of each year, so that updated records can be forwarded to the next teacher.

7.3 The school has a portfolio containing examples of children's work, matched to the level descriptions. All staff have discussed each piece of work and teachers' notes are attached, explaining how the assessment was made.

7.4 Teachers' own plans should indicate the focus for each unit of work and identify assessment opportunities.

## 8. RESOURCES

8.1 All children should have opportunities to use information technology including: *Junior Pinpoint, Datasweet, My World, 62 Honeypot Lane, Geosafari.*

At the upper end of the school, children will be given the opportunity to monitor the weather over a period of time using the school's weather monitoring station.

8.2 Children will have opportunities to use the following resources: globes, maps, atlases, aerial photographs, compasses, measuring equipment, cameras for recording.

8.3 Most geography equipment is kept within the central resources area. All equipment is readily accessible to the children. Children are given instructions in the safe and considerate use of resources, including taking care with consumables and materials which are not easy to store.

8.4 A full list of the resources available for geography have been set down by the geography co-ordinator. Curriculum equipment lists are attached to each policy as Appendix 2.

8.5 The geography co-ordinator is responsible for all geography resources, including ordering. The co-ordinator will ensure that the resource list is kept up to date.

## 9. EARLY YEARS

9.1 A sense of place is developed in children in their Reception year by activities which encourage the use of appropriate geographical language and a study of their immediate surroundings in the classroom, buildings and grounds.

9.2 The Reception teacher will plan appropriate activities in consultation with the Year 1 teacher, and use the school's Programme for Geography as a structure for identifying appropriate activities.

## 10. SAFETY AND CARE

10.1 When taking children out of school to undertake any study the school's Safety Policy should be adhered to fully.

10.2 When taking children on field trips or on out of county visits, the authority's guidelines should be adhered to fully. Teachers should be fully conversant with the authority's guidance on taking children on field trips before they begin planning or make any bookings.

10.3 When engaged in field work, children are required to display the same high standards of behaviour as those expected in school. They should behave in a considerate, responsible manner showing respect for other people and the environment.

All out of school activities will comply with the school's Health and Safety Policy. Copies of the policy can be found in the teachers' policy folder. Copies of all curriculum policies are available from the school secretary.

The school's usual Charging Policy applies to all out of school trips, including field visits. Details are set down in the school's Safety Policy and within the school prospectus. Copies of both documents can be found in the teachers' curriculum folder. Copies are available from the school secretary.

## 11. REVIEW

This policy is reviewed by the staff and governors in the Summer Term. Parents are most welcome to request copies of this document and comments are invited from anyone involved in the life of the school.

# Drawing up a Programme for Geography

## Mapping the geography curriculum

Alongside the Geography Policy, schools will need to map out a Programme for Geography which sets out clearly how the geography Statutory Orders are to be covered over the six years of primary education. This will aid teachers in their planning and will also enable them to monitor coverage.

Many factors, including policy decisions you have made, will contribute to the way you map out coverage in geography:

■ The cultural diversity within the school.

■ The number and type of school excursions, and the degree to which they form the basis of a geography study.

■ The term in which geography-based excursions take place.

■ The use of the schools' own facilities within geography projects, e.g. environmental areas.

■ Local events.

■ Whether geography is to be undertaken every term or if, in some terms, little geography will take place, thereby freeing time for a more in-depth project to be undertaken in another area of the curriculum.

■ Specific resources available.

■ Access to information technology equipment.

### An example Programme for Geography

**TASK 1: DRAWING UP THE OVERVIEW**

Many schools put a great deal of effort and time into implementing the previous Orders for geography. Schools should, therefore, take care not to make wholesale changes. Existing school plans should be incorporated, as far as is possible, into any new plans. The example in Figure 8 can be used as it stands, adapted to suit your requirements, or simply used as a guideline for whole school planning for geography. (A blank programme planning sheet – Figure 10 – can be found on page 33.) The example provides a minimum in order to meet the statutory requirements of the geography order. Schools may wish to add to this programme or to keep in their existing programme elements which are no longer part of the statutory requirements.

Key statements from the sample policy have been used to inform our decisions on when each aspect of geography is to be covered.

The example Programme for Geography covers the six years of primary education. It can be used by large and small schools as well as those schools with mixed aged classes. It can be used as a straightforward six-year delivery or as a two-year cycle. The first year of the cycle is shown as Year A and the second as Year B. It can be used by any school irrespective of the number of classes.

## TASK 2 – ENSURING COVERAGE

Task 2 takes the planning process a step further by setting down the Themes, Places and Skills that need to be addressed within each term. The numbers and letters used to identify the Places, Themes and Skills are those used in the new orders (and as shown on pages 8-11).

This detailed plan gives teachers more specific guidance to focus their planning and teaching.

## Comments on the example Programme for Geography

## KEY STAGE 1

The new Orders set down less geography to be covered at Key Stage 1. Where geography is set down on the Programme for a term it may be that the work is undertaken for only a part of that term.

At Key Stage 1, it is important to bear in mind that, by the end of the key stage, children should be developing:

■ A broader knowledge and understanding of the school locality.

■ A broader knowledge and understanding of at least one other locality, either in or beyond the UK.

■ A geographical vocabulary which they can use to describe a place.

■ An ability to make comparisons between places and geographical features.

■ An ability to use geographical skills such as observing, following directions, using and making maps, following routes on a map and identifying major features on maps and globes.

Whatever your organisation, at Key Stage 1 the Programme for Geography can be used just as it is set down.

The school locality study in the Autumn Term of each year has been written in two parts, in order to show progression in the work to be undertaken in this area. The Year 2 part of the Scheme of Work builds on the skills, knowledge and understanding of Year 1. However, teachers will find that it may be appropriate to include some of the work from the Year 2 programme with some of their more able Year 1 children. Also some of the Year 1 work could be more appropriate for less able Year 2 children. These units are very important, as all other geographical work during the two years will draw on the geographical skills, knowledge and understanding being developed during the school locality study. When drawing up more detailed plans for each of these two units of work, schools will therefore need to ensure there is progression within and across each unit of work.

Weather has been set down as an ongoing project throughout each term at Key Stage 1. It is a popular topic throughout primary schools, which are usually well equipped to undertake weather studies. Keeping records over time does not necessarily take up a great deal of curriculum time and yet the records can link valuably with work in other areas of the curriculum, such as mathematics.

The weather project is continued on the example programme into Key Stage 2, where the more complex ways that the weather can be recorded and data stored, and the links that can be made with the weather in other areas of the world can build well on work in Key Stage 1.

## KEY STAGE 2

It was generally accepted that the old Orders presented too great a workload for teachers, and demanded far too much curriculum time. Schools should therefore ensure that they are now clear what, in the new orders, is to be taught within each year so that the teachers' own plans do not collectively present too great a workload at Key Stage 2.

At Key Stage 2, it is important to bear in mind that, by the end of the Key Stage, children should be developing:

■ A broader knowledge and understanding of places and themes.

■ A greater ability to describe what places are like, using geographical vocabulary.

■ An ability to offer explanations for certain characteristics of places and recognise patterns.

■ An ability to identify physical and human processes and describe some of their effects upon the environment.

■ An ability to gather information from secondary sources as well as from first hand observation.

■ An ability to use geographical skills, such as: observing and communicating using geographical vocabulary, measuring and recording, interpreting information from maps, using maps and photographs, making and using their own maps and using atlases.

The school locality studies undertaken at the start of Years 3 and 4 each have a different theme. As in Key Stage 1, subsequent geographical work during the following years will draw on the concepts and skills being developed during the school locality study. Detailed plans for these units of work must, therefore, ensure a progression within and across each study.

The Summer Term for Year 5 and 6 classes is popular for undertaking fieldtrips to a contrasting locality in the UK. The project 'Forests' has been chosen as an ideal study as it presents a superb opportunity to develop the environmental theme. There is also a wealth of support material available on the theme of forests, much of which is listed in Section 7. It also offers opportunities for linking geography work with other curriculum areas, e.g. science. Teachers may decide to undertake an alternative project to 'Forests' during this term, but will need to ensure that the Places, Themes and Skills that form the focus for the project are the same as those set down on the Programme for Geography.

Although the work planned for each term can be moved to another term to suit individual schools' own planning, some projects may well be better undertaken in certain terms. The Summer Term usually provides more reliable and suitable weather for field work, particularly for plans beyond the school grounds, allowing classes to get outside more to observe in their locality and other areas at first hand. This term is, however, an 'action packed' one for primary schools, with many different events taking place. (The Summer Term should be considered, for example, as the time to make summative assessments of children's work, and to prepare reports for the parents – particularly with Year 2 and Year 6 children.)

The project set down on the Programme for Key Stage 2 called 'Here, there and everywhere' focuses on building up an awareness of the world extending beyond their own locality. This allows children to draw on all their previous experiences, and to develop a greater understanding of both the great diversity yet interdependency of our world. Referring to maps and globes and talking about current issues are activities commonly undertaken by primary schools, and schools may be able to continue this aspect of geography without having to make too many changes to their existing plans.

This example Programme for Geography forms the basis of the Scheme of Work which follows in Section 5.

# Figure 8: An example Programme for Geography – overview



**PROGRAMME FOR GEOGRAPHY (TASK 1 – THE OVERVIEW)**

| | | | | | | | |
|---|---|---|---|---|---|---|---|
| | YEARS | 1 | 2 | 3 | 4 | 5 | 6 |
| AUTUMN | ONGOING STUDY – WEATHER | The locality of the school<br>– our school | The locality of the school<br>– our school | The locality of the school<br>– our place | The locality of the school<br>– our place | Thematic Study<br>– a named river | Thematic Study<br>– a named issue |
| SPRING | ONGOING STUDY – WEATHER | | | Weather | | | Here, there and everywhere |
| SUMMER | ONGOING STUDY – WEATHER | Thematic Study<br>– a named locality | Contrasting locality overseas<br>– Kaptalamwa in Kenya | | Locality in South or Central America, Africa or Asia<br>– Kaptalamwa in Kenya | Contrasting locality in the UK<br>– a named forest | |
| | CYCLE | A | B | A | B | A | B |

# Figure 9: An example Programme for Geography – ensuring coverage

**PROGRAMME FOR GEOGRAPHY (TASK 2– ENSURING COVERAGE)**

| YEARS | 1 | 2 | 3 | 4 | 5 | 6 | |
|---|---|---|---|---|---|---|---|
| **AUTUMN** | 1a,b<br>Skills: 2; 3a,b,c,d,e<br>Places: 4; 5a,d | 1a,b<br>Skills: 2;3a,b,c,d,e<br>Places: 4; 5a,d | 1a,b,c,d<br>Skills: 2a,b,c 3a,b,c,d,e,f<br>Places: 4; 5a,d,e<br>Themes: 6; 9b | 1a,b,c,d<br>Skills: 2a,b,c 3a,b,c,d,e,f<br>Places: 4; 5a,b,c,e<br>Themes: 6; 9a | 1a,b,d<br>Skills: 2a,b,c 3a,b,c,d,e,f<br>Places: 5; a,e<br>Themes: 6; 7a,b | 1a,b,d<br>Skills: 2a,b,c 3a,c,d<br>Places: 4; 5a,c,d,e<br>Themes: 6; 9c | |
| **SPRING** | | | 1a,b,c,d<br>Skills: 2a,b,c 3a,b,d,e,f<br>Places: 4; 5b,e<br>Themes: 6; 8a,b,c | | | 1a,b,c,d<br>Skills: 2a,b,c 3d,e,f<br>Places: 5b,e<br>Themes: 6 | |
| **SUMMER** | 1a,b,c<br>Skills: 2; 3a,c,d,e<br>Places: 5a,b,d<br>Themes: 6a,b,c | 1b,c<br>Skills: 2; 3a,e,f<br>Places: 4; 5a,b,d | | 1a,b,c,d<br>Skills: 2a,b,c 3a,d,e<br>Places: 4; 5a,b,c,e<br>Themes: 6; 9a,b | 1a,b,d<br>Skills: 2a,b,c 3a,b,c,d,e,f<br>Places: 4; 5a,b,c,d,e<br>Themes: 6; 10a,b | | |
| **CYCLE** | A | B | A | B | A | B | |

1a,b SKILLS: 2; 3a,b,c,d,f PLACES: 4; 5c

32

# Figure 10: A Programme for Geography – blank template

PROGRAMME FOR GEOGRAPHY

| YEARS | 1 | 2 | 3 | 4 | 5 | 6 |
|---|---|---|---|---|---|---|
| AUTUMN | | | | | | |
| SPRING | | | | | | |
| SUMMER | | | | | | |
| CYCLE | A | B | A | B | A | B |

# 5 A Scheme of Work

The aim of this section is to provide a Scheme of Work for geography at Key Stages 1 and 2. However, as every school's local area is unique, it is not possible to provide a Scheme of Work which is completely relevant to each school. Some adjustments may have to be made where features mentioned do not exist in your area. This may mean that, if adjustments are made, then some programmes of study may not be covered. Equally, it may be that other programmes of study may be touched upon, but not covered in depth or detail. The suggested activities are, however, flexible enough to provide the guidance needed to successfully deliver the National Curriculum for geography.

For ease of reference, the activities which make up this Scheme of Work are set out as double-page, photocopiable spreads, with the focus programmes of study for each activity clearly indicated.

The Scheme of Work includes activities planned for a contrasting locality outside the UK at Key Stage 1 and a locality in a country in South or Central America, Asia (excluding Japan) or Africa at Key Stage 2. For the purpose of this Scheme of Work, we have used the Geographical Association's photopack *Kaptalamwa, A Village in Kenya*. The photopack contains many further activities to extend and develop the study of a distant place for those schools who wish to do so.

There are three possible ways in which schools may choose to use this Scheme of Work:

1. Use the example Programme of Geography and the accompanying Scheme of Work as it stands – knowing that it meets all the statutory requirements for geography at Key Stages 1 and 2.

2. Adapt the example Programme for Geography and the Scheme of Work to suit your existing plans and resources.

3. Use the Programme for Geography and the Scheme of Work simply to identify the tasks that are to be undertaken, and use our guidance in order to produce your own unique Programme for Geography and associated Scheme of Work.

There is a photocopiable, blank planning sheet for the Programme for Geography on page 33.

**N.B.** *To identify elements of the National Curriculum Orders referred to in the unit plans, see our summary of the requirements in Figures 3 and 4 (pages 8-11).*

# YEARS
# 1 & 2

## Title
Our school

## Year
1: Autumn Term

## Cycle
A and B

## Place
The locality of the school

| PROGRAMME FOR GEOGRAPHY (TASK 1 THE OVERVIEW) | | | | | | |
|---|---|---|---|---|---|---|
| Y | 1 | 2 | 3 | 4 | 5 | 6 |
| ONGOING STUDY – WEATHER | The locality of the school | The locality of the school | The locality of the school | The locality of the school | Thematic Study | Thematic Study |
| | | | Weather | | | Here, there and everywhere |
| | Thematic Study | Contrasting locality overseas | | Locality in South or Central America, Asia or Africa | Contrasting locality in the UK | |
| | A | B | A | B | A | B |

## Focus
Looking at our school and grounds

## National Curriculum
1a,b
2
3a,b,c,d,e
4
5a,d

## Useful resources
A simple large-scale plan of the school site. Photographs of features in the school locality.

## Suggested activities

### STARTING POINT

A walk around the school site to look at and discuss the features of the site.

### MAKING OBSERVATIONS

What buildings are on the site? How many buildings are there altogether? What is each building used for? How can you tell? Who uses the buildings? How many storeys does each have? How old are the buildings (new, quite old, very old)? Have there been any additions to the buildings? How can you tell? If you have had recent additions to the buildings, ask the children why they think they were needed. Are there any unused buildings? Why are they unused now? Children should have opportunities to discuss the similarities and differences between the various buildings/parts of buildings on site.

What other built features can the children identify (e.g. fences, walls, a car park, a playground)? What is the purpose of each feature? Why are they needed/Who uses them? Encourage the children to use locational language, such as 'in front of', 'near to' and so on, to describe the position of the features.

### IN THE CLASSROOM

Individual or small groups of children can select a particular feature of the school's locality about which to record information. This could contribute to a class display about "Our School" centred on a large-scale plan. (This might include "our favourite places" in the school and grounds.)

**MAKING RECORDINGS**

Take a number of photographs as you stop at various points around the school. These can form a display in the classroom. The display can be usedfor a short discussion with the children to give them the opportunity to relate the photographs to features they observed earlier.

**USING MAPS**

Introduce the name of the locality to the children. Which children live in this locality and which live in another place? The children could record this information on a class graph. Some children could extend this to find out where the children in other classes in the school live. Children from another locality should have opportunity to talk about what it is like where they live. Is everywhere the same? Encourage them to think about any similarities as well as the differences.

## Title
Our school

## Year
2: Autumn Term

## Cycle
A and B

## Place
The locality of the school

| PROGRAMME FOR GEOGRAPHY (TASK 1 THE OVERVIEW) | | | | | | |
|---|---|---|---|---|---|---|
| Y | 1 | 2 | 3 | 4 | 5 | 6 |
| ONGOING STUDY – WEATHER | The locality of the school | The locality of the school | The locality of the school | The locality of the school | Thematic Study | Thematic Study |
| | | | Weather | | | Here, there and everywhere |
| | Thematic Study | Contrasting locality overseas | | Locality in South or Central America, Asia or Africa | Contrasting locality in the UK | |
| | A | B | A | B | A | B |

## Focus
From our school to beyond – what is it like here?

## National Curriculum
2
3a,b,c,d,e
4
5a,b

## Useful resources
A simple teacher-drawn map of the area to be used for field work. Large-scale map of area (1:1250 or 1:2500). Compasses.

## Suggested activities

### STARTING POINT

A walk outside, to consider the school site and beyond.

### MAKING OBSERVATIONS

Consider the ground which the school site occupies. How much ground space does each of the built features on the school site occupy (is it small/large)? Discuss, if relevant, why the feature is sited in that location. Examine any 'natural' features there, e.g. soil, water. Where can these be seen? Is the ground flat? Are there any slopes? If so, where are the slopes? If your school has a grassed area, encourage the children to think about the amount of space taken up by this, compared to the buildings and other features. Allow children to express how they feel about the school locality.

Children who are familiar with the features of the school site should have a simple plan of the school grounds for this activity, with the boundary, buildings and playground marked. The plan should also have North indicated. This will allow them to follow the route and identify where other, smaller features are located. The children can mark the position of these on their plan, using pictures or symbols.

Look at the land outside the grounds from several different points on the school site. Is this flat/sloping/very hilly? Is there a difference between the viewpoints? What different features can the children identify from these viewpoints around the school, e.g. houses, shops, road, pavement, hill, field? Encourage children to describe where a particular feature is as well as pointing to it! If necessary, identify a few unfamiliar features for the children, e.g. factory, grass verge, stream, and describe the location. Introduce the direction of view, i.e. north, south, east, west.

## IN THE CLASSROOM

Introduce the large-scale map with their school on to the more able children. Which of the features on the map have they seen? Encourage the children to use directional language in describing the route and location of features. Small groups of children could select a particular type of land use, or the land use on a particular street about which to record information. Some children might want to record their evidence about the type of work adults seen were involved in, or what features they were able to identify on their route.

## USEFUL TIPS

Paint a large, eight-pointed compass rose on your school playground so that children can relate aspects of the school to direction. The compass can also be used during PE activities to indicate the direction children should move in.

Extend the work to the area surrounding the school within easy access. Ideally this will involve direct experience for the children, but if this is not possible, and to extend this study, then large-size photographs showing some of the features of the locality can be used. Ensure that photographs include the 'life' of the locality and some views as well as individual features. What different uses of land are there? Which uses of land/activities take up more space? What work do people do in your locality? What local landscape features are there? What do they find attractive/unattractive about where they live?

Children who have followed a route on the school grounds map in Year 1 could follow the route taken during their field visit on a simple map of the area. Encourage them to orientate their map correctly by matching the map to the ground features and to a compass.

## Title

**Nearby named locality** (ideally a village if your locality is not part of a village)

## Year

1/2: Summer Term

## Cycle

A

| PROGRAMME FOR GEOGRAPHY (TASK 1 THE OVERVIEW) | | | | | | |
|---|---|---|---|---|---|---|
| Y | 1 | 2 | 3 | 4 | 5 | 6 |
| ONGOING STUDY – WEATHER | The locality of the school | The locality of the school | The locality of the school | The locality of the school | Thematic Study | Thematic Study |
| | | | Weather | | | Here, there and everywhere |
| | Thematic Study | Contrasting locality overseas | | Locality in South or Central America, Asia or Africa | Contrasting locality in the UK | |
| | A | B | A | B | A | B |

## Place

Named contrasting locality in the UK for a thematic study

## Focus

The quality of the environment

## National Curriculum

1a,b,c
2
3c,d,e
5a,b,d
6a,b,c

## Useful resources

A simple map of the locality. A large-size map of the UK (template 6 on page 106). Photographs for follow up work. Road atlas.

## Suggested activities

### STARTING POINT

A discussion about the visit, including safety issues, leading to field work at the locality.

### MAKING OBSERVATIONS

Begin by drawing children's attention to a change happening in the locality, (e.g. new building work/demolition, roadworks). What changes can the children see? Initially children may suggest changes linked with the seasons. Which changes are made by people? Who is making the changes? Who will the changes affect? Which changes do the children think will make the environment better? Does everyone agree? Discuss what we might mean by 'better'. Discuss with the children what other improvements might be made to the locality. Who would benefit?

Discuss the physical environment of any built-up place too. What flat places are there in the locality (e.g. car park, games pitch)? What slopes are there? If the locality itself is on very flat land then the journey to it may afford opportunities for discussing 'ups and downs' (e.g. bridges, steps, hills, embankments). Introduce relief words (e.g. highest, lowest, steep, flat, uneven, level, hill, hillside, hilltop) where appropriate. Is there a stream or river in the locality? If so, where? What is the ground surface like?

What type of homes are there? Houses/bungalows? One/two/more storeys? On their own or joined to others (detached, semi-detached, terraced)? What materials are they made from? Choose one to examine carefully. Look at wall materials, roof materials, boundaries. How old is the house (new, old, very old)?

What else is there to see on your route around the locality besides homes (e.g. church, school, shop, playground, field)? Who uses these? How are each of these used? Select at least one feature to consider in more detail (ideally, a shop) and compare it with a similar feature in your locality. What do the children find attractive/unattractive about it? How is it similar to and different from their locality?

The names of some of the roads, particularly old established roads, can prove interesting to children e.g. Pond Lane, North Street, Church Lane. Encourage the children to suggest why the roads have these names and look for any evidence in support of their suggestion. (This may also provide evidence of change.)

Which roads are busy/not busy? What kinds of transport are being used in/through the locality? Children could undertake a simple, short traffic survey for a fixed time if there are safe, convenient points.

### IN THE CLASSROOM

Where were the changes happening? Children can colour where changes are happening on a simple map of the locality, and photographs of changes which have been identified can be matched to a large-scale map back in the classroom.

Children could draw or paint pictures of the houses and other features of the village to link with their large size wall map of the village (or part of the village, if necessary).

### USING MAPS

Locate the village on the map of the UK for the children to compare it with their own locality. Which direction did they travel to get to the village? Draw maps of the journey, showing some of the features they passed along the way. Some children may enjoy trying to follow the route on a large-scale road map. The map should include pictorial symbols to help identify some of the distinctive features and their location. Children may add further features to their map, showing where changes are taking place.

### USEFUL TIP

Ensure the recording sheet is a simple tick sheet if there is a fairly regular traffic flow.

### CROSS-CURRICULAR LINKS

Science – Sc3 Materials and their properties.

**Title** Kaptalamwa in Kenya

**Year** 1/2: Summer Term

**Cycle** B

**Place** Contrasting locality overseas

| PROGRAMME FOR GEOGRAPHY (TASK 1 THE OVERVIEW) | | | | | | |
|---|---|---|---|---|---|---|
| Y | 1 | 2 | 3 | 4 | 5 | 6 |
| ONGOING STUDY – WEATHER | The locality of the school | The locality of the school | The locality of the school | The locality of the school | Thematic Study | Thematic Study |
| | | | Weather | | | Here, there and everywhere |
| | Thematic Study | Contrasting locality overseas | | Locality in South or Central America, Asia or Africa | Contrasting locality in the UK | |
| | A | B | A | B | A | B |

**Focus** The similarities and differences between Kaptalamwa and our school locality

**National Curriculum**
1b,c
2
3a,e,f
4
5a,b,c

**Useful resources** The Kaptalamwa pack, including photographs from the pack, showing people, scenery, homes, school and transport. A large-scale map-drawing grid (template 1 on page 101). Photographs of your locality. A globe. A simple atlas. A 'Junior' atlas. Travel brochures.

## Suggested activities

**STARTING POINT**

Photograph 11 from the Kaptalamwa pack.

**IN THE CLASSROOM**

Look closely at photograph 11. What are the people in the photograph doing? How can we tell? Who do they think is with the children? How many children are there in this family? What do the clothes and eating outside tell us about the weather there? What are the family sitting on? If necessary point out that the hut that the children sleep in is in the photograph. What else can they identify in the photograph (fence, trees, soil, chicken, bowls, water can, storage hut for potatoes or sweetcorn)? Which of these things do we have/use in our locality?

Use photograph 12 (Joseph's hut). What is the hut like? Describe its shape, size, colour, building materials. How is it different from the hut in the first photograph (newly built, not finished)? How does it compare with their own homes?

Look at photographs 1, 2 and 17. Ask the children to describe the landscape. (Children will be able to identify more features by using a magnifying glass.) Use the large grid over the photographs to identify particular parts of the photograph and stimulate enquiry, e.g. What can you see in square B4? In which squares can you see a hut? Encourage the use of geographical language, such as 'hilly', 'woodland' and so on. Which of these things are there in our locality? [Note: individual shambas – small farms – can usually be identified on the photographs by the small patches of trees.]

Using photograph 17, say how many buildings there are in a shamba (usually 1 'house', 2 or 3 living huts and 2 or 3 storage huts for potatoes or sweetcorn). Use the plan of the Malakwen shamba on page 1.25 as the basis for a model shamba.

Extend the activity, using photographs which show people doing various activities – particularly 3, 4, 5, 7, 10, 13, 14 and 15. How do the people usually spend their day? (You will need to explain what is happening in some of the photographs.) Start with the activities in the huts and in the shambas (i.e. at home), then extend the discussion to include the village centre (i.e. the locality). In each case give the children time to identify familiar items or aspects of life, encourage them to suggest explanations for what they observe, and ask them to think about the similarities/differences with their locality. Compare Jelimo Malakwen's day (see page 1.26) with their day on a time line.

Finally, the children can use a globe and simple atlas to identify the UK first, then Africa. You may need a 'junior' atlas to find Kenya. Use the map in the pack to locate Kaptalamwa. Discuss with the children the directions involved. Use travel brochures and photographs 21, 22, 23 and 24 to show that not all of Kenya is like Kaptalamwa.

**CROSS-CURRICULAR LINKS**

Art – making a model shamba.

## Title
Weather

## Year
1/2: Ongoing study

## Cycle
A and B

## Place
The locality of the school

| | PROGRAMME FOR GEOGRAPHY (TASK 1 THE OVERVIEW) | | | | | |
|---|---|---|---|---|---|---|
| Y | 1 | 2 | 3 | 4 | 5 | 6 |
| ONGOING STUDY – WEATHER | The locality of the school | The locality of the school | The locality of the school | The locality of the school | Thematic Study | Thematic Study |
| | | | Weather | | | Here, there and everywhere |
| | Thematic Study | Contrasting locality overseas | | Locality in South or Central America, Asia or Africa | Contrasting locality in the UK | |
| | A | B | A | B | A | B |

## Focus
The effects of our weather

## National Curriculum
1a,b
2
3a,b,c,d,f
4
5c

## Useful resources
Simple weather measuring equipment (details of where to obtain such equipment is given in Section 7 of this book). Weather recording sheet (templates 4 and 5 on pages 104 and 105). *Time and Place* Photopack. Key Stage 1 non-statutory standard assessment task for technology entitled *Protective Headgear*. The pack was originally distributed to all primary schools in England and Wales. A simple map of your school grounds.

## Suggested activities

### STARTING POINT

At least three occasions with different weather conditions in the course of school year, e.g. windy day, hot day, very cold/frosty or snowy day, rainy day.

### MAKING OBSERVATIONS

What effect has the weather outside had on us inside the classroom (e.g. it feels colder/warmer, we need to put the lights on, the heating has to be turned down/up/off)? What is the weather like outside? (This will give you the opportunity to develop the children's weather vocabulary.)

Consider the effects on the school buildings and grounds. Look carefully at any woodwork on the building (e.g. peeling paint), soil areas (e.g. stains through splashes on to buildings, cracks in the soil), dips in the ground (e.g. puddles forming), windy corners where litter/leaves collect. Does the weather have the same effect on all parts of the school – buildings and grounds? (Use a simple map of the school grounds to plot where the effects are found.)

What other weather effects can the children detect/do they know about from travelling to school? Discuss the effect the weather has on road transport. Does the weather affect the way any of their family chooses to travel to work or school?

Provide opportunities for children to consider the longer term effects of a spell of very similar weather, e.g. warm, dry conditions over a long time may lead to plants dying, water shortages. Photographs, such as those in *Time and Place* Photopack, or other suitable large clear photographs can be used to enhance this part of the study. Extend the activity by using books about weather conditions and seasons, particularly for the UK.

When children are investigating other localities, the weather experienced there can be discussed using evidence from appropriate photographs.

## IN THE CLASSROOM

As the weather doesn't come to order you will need to plan this study carefully – ideally devoting at least part of one day each term to the it, but with enough flexibility in your programme to 'slot it in' when the weather permits.

The framework for studying each type of weather will be similar. In each case, consider what effects the weather has on the clothing we choose to wear, including footwear and headgear. Younger children could decide how best to dress their teddy/doll for these weather conditions. The children could undertake the Design and Technology non-statutory task on hats, which was supplied free to all primary schools.

It is envisaged that the weather project will be undertaken over a period of time. Children should be given opportunities to record the weather over a period, using the weather recording sheet for KS1, and observe the changes and similarities in their recordings. This work can be developed further at Key Stage 2, using more sophisticated equipment, techniques, and recording methods.

A weather project will also provide an ideal opportunity to help the children to learn more about symbols. They should be encouraged to design their own, as well as to recognise some more well-known weather symbols (e.g. those used by the BBC).

## CROSS-CURRICULAR LINKS

Science – observing warm and cold conditions and their effect on people. How the properties of materials affect what we wear in different weather conditions.

English.

Design and Technology – designing and making hats for particular weather conditions.

# YEARS
# 3 & 4

## Title
**Our place**

## Year
3/4: Autumn Term

## Cycle
A

## Place
The locality of the school

## Focus
Land use in our locality

## National Curriculum
1a,b,c,d
2a,b,c
3a,b,c,d,e,f
4
5a,d,e
6
9b

| PROGRAMME FOR GEOGRAPHY (TASK 1 THE OVERVIEW) | | | | | | |
|---|---|---|---|---|---|---|
| Y | 1 | 2 | 3 | 4 | 5 | 6 |
| ONGOING STUDY – WEATHER | The locality of the school | The locality of the school | The locality of the school | The locality of the school | Thematic Study | Thematic Study |
| | | | Weather | | | Here, there and everywhere |
| | Thematic Study | Contrasting locality overseas | | Locality in South or Central America, Asia or Africa | Contrasting locality in the UK | |
| | A | B | A | B | A | B |

## Useful resources
A large-scale map of area (1:1250 or 1:2500). Oblique aerial photographs of your local area. Medium-scale (2cm squared) grids (template 2 on page 102). Magnifying glasses. Landscape and townscape photographs of the UK.

N.B. Teachers will note that a smaller-scale map grid has also been provided for use where necessary (template 3 on page 103).

## Suggested activities

### STARTING POINT

A short walk in the school locality, which will provide opportunities to consider different uses of land.

### IN THE CLASSROOM

Use oblique aerial photographs of the local area. These will allow children to see types of land use not always visible from the ground and often cover a larger area of ground than can be covered on foot. In what ways is the land being used? Make a list of all the ways in which land is being used in your area. Let the children think of ways in which the list can be grouped. You may need to introduce the appropriate geographical vocabulary for the groupings at this point, e.g. leisure, industry. The children can suggest who might use this land – Which are 'natural' uses of the land and which uses include human features?

### USING MAPS

Children could follow the route of the walk for themselves on a large-scale map, or the route may be marked on the map for them if necessary. The different types of land use alongside the route can be indicated on the map. Are there any recent or proposed changes to the land usage?

Compare the aerial photographs with the local map at the 1:2500 scale. What further information does this provide? Encourage the children to consider built up and open areas. Using a simple teacher-drawn outline map of the local area the children can now map the land use. Discuss with the children how they will ensure that their map is clear. Which groupings will they use? Will the class agree a set of colours/symbols?

Cover the map with a clear 2cm grid square and use the map to answer questions about your area. What is the main type of land use in your area? How much of the land is occupied by the main land use (count the grid squares)? Is yours mainly a rural or urban area? How much variety of land use is there? Where are particular types of land use located (use co-ordinates)? Discuss with the children possible reasons for particular land uses in their area (start with the reason for their school).

Extend the work to discuss the ways in which land is being used by showing a selection of any good quality landscape and townscape photographs of the UK – either aerial or views. Try to include some land uses which are not part of the local area. Which places have similar land uses to their own area? What unfamiliar land uses can they identify? Into which of their groupings would these land uses fit? What new groupings will they need? Encourage the children to suggest the possible reasons for the differences in land use in other places. Identify where the places are on atlas maps of the UK. Compare with physical and other maps of the UK to consider what patterns of land use exist. Encourage the children to provide their own explanations for any patterns identified.

## USEFUL TIPS

Get the children to think about the use of symbols to represent various types of land use beforehand, ready to mark on their local map. If there are a large number of types of land use in your locality, then different groups of children could record the particular ones. Does land use vary much along each street? Which parts of your locality provide a large number of land use types/only one type of land use?

A magnifying glass will help the children to identify features on the photographs.

| | Title | Our place |
|---|---|---|


**Title**  Our place

**Year**  3/4: Autumn Term

**Cycle**  B

**Place**  The locality of the school

**Focus**  The character of our locality

**National Curriculum**
1a,b,c,d
2a,b,c
3a,b,c,d,e,f
4
5a,b,c,e
6
9a

| PROGRAMME FOR GEOGRAPHY (TASK 1 THE OVERVIEW) | | | | | | |
|---|---|---|---|---|---|---|
| Y | 1 | 2 | 3 | 4 | 5 | 6 |
| ONGOING STUDY – WEATHER | The locality of the school | The locality of the school | The locality of the school | The locality of the school | Thematic Study | Thematic Study |
| | | | Weather | | | Here, there and everywhere |
| | Thematic Study | Contrasting locality overseas | | Locality in South or Central America, Asia or Africa | Contrasting locality in the UK | |
| | A | B | A | B | A | B |

**Useful resources**  Scale maps of your area at a variety of scales (including 1:1250 or 1:2500, 1: 25 000 and 1:50 000). Information collected about the school locality and its wider area. Aerial photographs of your region.

**Suggested activities**

STARTING POINT

A walk in the school locality (or clear photographs showing main characteristics of the school's locality).

IN THE CLASSROOM

What kinds of economic activities can be seen/is there evidence for along the route? What variety is there? Where are the economic activities of your locality sited? Are they well spaced or grouped together? For whom do the economic activities provide employment? What communication links does each have? Consider whether the children could interview or write to employers to get more information about the economic activities in your area.

Discuss what links your settlement has with others around it; e.g. conduct a class shopping survey – where do the children go to buy food, clothes, toys? What reasons are there for going to another settlement? How do/can they travel to that settlement?

What evidence does information collected about the school locality and its wider area provide about what type of settlement the school is in (village, town, suburb, city)? What further evidence is there of what their settlement is like and how it has changed? Do the street names provide any clues to former functions?

USING MAPS

If possible, take the children out into their locality to identify what characteristics it has. Encourage them to think about both physical and human features and mark these on a simple, large-scale map. Children could follow the route for themselves on the map, or the route may be marked on the map for them if necessary.

Encourage the children to describe the location of their settlement in relation to landscape (e.g. river, hills, farmland, coast) and routes to their settlement, using the local map to help them. Draw a map of your settlement (or part of it) to show its locational features.

Using a 1:25 000 or 1:50 000 map, which includes your settlement, identify the built up areas. Discuss with the children which are villages, towns, etc. How do we know? Is it always possible to tell from the map? Compare the sizes of the settlements on the map. Children could use tracing and squared paper to draw around the built up areas to compare their areas more precisely. Are all villages similar in size? Are the villages always smaller than towns? For comparison, include a map of the closest city (if you are in a rural area) or rural area (if you are in an inner city or the suburbs).

The use of aerial photographs could extend the work further to include other parts of the locality not covered by the survey or nearby areas. Photographs of particular economic activities can be matched by location to a large-scale local map. (Use atlas maps of the UK to locate the places shown in the photographs.)

Small groups of children could research one nearby settlement, focusing on its size, location and characteristics by using, for example, appropriate maps, tourist information and photographs and report back to the rest of the class.

## INFORMATION TECHNOLOGY

Back in the classroom, a simple database could be used to store the information gathered. This should provide opportunities to consider whether there is any pattern to the siting of economic activities. Ask the children to suggest reasons for their location.

## CROSS-CURRICULAR LINKS

History – local study.

| | **Title** | Weather |

| | **Year** | 3/4: Spring Term |

| | **Cycle** | A |

| | **Place** | Various named places |

| | **Focus** | Differences in the weather here and elsewhere |

| | **National Curriculum** | 1a,b,c,d<br>2a,b,c<br>3a,b,d,e,f<br>4<br>5b,e<br>6<br>8a,b,c |

| | **Useful resources** | Map of the school grounds. Thermometer, rain gauge, anemometer and wind sock (or equivalent). Weather recording sheet for Key Stage 2 (template 5 on page 105). Compass. National newspapers with weather information for the period of the topic. Books for children about the weather in parts of the world with very different weather conditions. |

**PROGRAMME FOR GEOGRAPHY (TASK 1 THE OVERVIEW)**

| Y | 1 | 2 | 3 | 4 | 5 | 6 |
|---|---|---|---|---|---|---|
| ONGOING STUDY – WEATHER | The locality of the school | The locality of the school | The locality of the school | The locality of the school | Thematic Study | Thematic Study |
| | | | Weather | | | Here, there and everywhere |
| | Thematic Study | Contrasting locality overseas | | Locality in South or Central America, Asia or Africa | Contrasting locality in the UK | |
| | A | B | A | B | A | B |

**Suggested activities**

**STARTING POINT**

A question: Is the weather the same all over our school grounds?

**IN THE CLASSROOM**

Encourage the children to think about what differences they have noticed in the playground (e.g. windy in a particular corner). Where might differences occur and why? Consider temperature, rainfall, wind speed and direction. Take the children into the grounds to decide on possible locations to collect evidence. These should ideally include North and South facing sites (temperature), exposed and sheltered sites (rainfall, temperature, wind speed and direction). Mark these clearly on a map of the grounds. Some children will also need these locations marked on the ground.

Use the information in the newspaper (or on Ceefax or Teletext) to find out what the weather is like in other parts of the world. Select at least three places in very different situations, including UK, Europe and elsewhere in the world, e.g. Reykjavik, Rome, Kathmandu, Bangkok. Use an atlas to locate the named places. The children should describe the location of the places in the world in relation to such features as the equator, poles, coasts, seas, mountains and continents. Collect weather information for each for about four weeks (usually temperature and weather type is provided). What picture emerges of what the weather is like in each place? How much variation is there in the weather for each during the time covered? Use reference books to find out more about the weather in these parts of the world and how it affects the lives of the people there.

Discuss the pattern of the weather identified in *The House on the Hill* by Philippe Dupasquier and in poetry relating to the seasons. What sort of weather do we expect to have in the different seasons? Is the weather always/usually what we expect within the seasons? How might it vary? A taped daily forecast from the television for one day from each season would be useful to promote further discussion.

## USING INFORMATION TECHNOLOGY

The information could be transferred to a suitable database using a program such as *Junior Pinpoint* on an Archimedes computer. The records should be continued for a few weeks, including a spell of sunny weather and windy weather to allow clearer analysis of the data.

## ANALYSING THE DATA

What can the children tell you about the weather around the school from looking carefully at the data collected? What evidence is there to support their responses?

From which direction does the wind most frequently blow (exposed site)?

How is the wind speed and direction affected by a sheltered site?

How is the temperature affected by the direction a site is facing (its 'aspect')?

What link is there between the amount of cloud cover and rainfall?

What effect does an exposed/sheltered site have on the weather?

What explanations can the children provide to account for their results?

What patterns have they found?

Compare the data collected with the data provided in the newspaper for your area. How accurate is the information provided? What might be the reasons for any inaccuracy?

## MEASURING AND RECORDING

Children should carefully measure and record the temperature, rainfall, wind speed and direction every day at the same time (e.g. just before lunchtime). Include a note of the amount of cloud cover. A photocopiable recording sheet is provided (template 5 on page 105).

## CROSS-CURRICULAR LINKS

English.

| | **Title** | Kaptalamwa in Kenya |
|---|---|---|

**Title** Kaptalamwa in Kenya

**Year** 3/4: Spring Term

**Cycle** B

**Place** A locality in a country in Central or South America, Africa or Asia (except Japan)

| PROGRAMME FOR GEOGRAPHY (TASK 1 THE OVERVIEW) | | | | | | |
|---|---|---|---|---|---|---|
| Y | 1 | 2 | 3 | 4 | 5 | 6 |
| ONGOING STUDY – WEATHER | The locality of the school | The locality of the school | The locality of the school | The locality of the school | Thematic Study | Thematic Study |
| | | | Weather | | | Here, there and everywhere |
| | Thematic Study | Contrasting locality overseas | | Locality in South or Central America, Asia or Africa | Contrasting locality in the UK | |
| | A | B | A | B | A | B |

**Focus** How the features of Kaptalamwa affect life there

**National Curriculum**
1a,b,c,d
2a,b,c
3a,d,e
4
5a,b,c,e
6
8c
9a,b

**Useful resources** Kaptalamwa photopack. Travel brochures of Kenya. Books on Kenya. Atlases. 1:25 000 local map or road atlas.

**Suggested activities**

**STARTING POINT**

A display focusing on the Kaptalamwa booklet, showing huts and houses, food crops and water. Information from the Kaptalamwa booklet. Travel brochures of Kenya.

**IN THE CLASSROOM**

Discuss with the children what is essential for life: food, water and shelter. Which things that the Marakwet people need are available on and around their own shamba? How do the Marakwet people use their land to provide some of the things they need? Introduce the terms self-sufficient and cash-crop (pyrethrum).

The children could draw a simple landscape sketch using one of the Kaptalamwa landscape photographs, colouring the different types of land use or annotating the sketch. Some children will need you to provide an outline of the sketch for them to complete.

What do the crops and scenery indicate about the type of weather experienced in Kaptalamwa? Use the weather statistics from the photopack to compare with the sort of weather you experience. Look at the location of Kaptalamwa and the Cherangani Hills in the photopack. Compare with your location. Look at the pictures and photographs of scenery and weather in Kenya shown in travel brochures. How is it different? The children may be able to suggest some reasons for the differences. They may also need to do further research using the information provided in the pack.

Use the photographs of Kaptalamwa village centre and information booklet to see what goods can be obtained there. Which services are also available?

Make lists comparing Kaptalamwa to your locality. For which of the goods and services available locally in Kaptalamwa do you need to travel outside your locality – i.e. where a vehicle is required? (Children living in small villages may be surprised at the result.)

Now use the photographs of Kitale and Eldoret. What additional goods and services can you identify from the photographs? Find further information by using the Kaptalamwa booklet. What do people in Kaptalamwa need to buy in Kitale or Eldoret? Which services are available here? Which are 'essential' and which 'luxuries'? This may promote some interesting debate amongst the children! What other reasons might there be for travelling to other places/larger places? Where do the children and their families go for a wider range of goods and services? Use a 1:25 000 map (or road atlas if necessary) to measure the distance and direction to these places. Compare with the situation in Kaptalamwa using the Cherangani Hills map.

## USING MAPS

Use atlases to find the names and locations of the main towns of Kenya. What kind of additional facilities might be available here (e.g. tourism in Nairobi and Mombasa)? Can the children find any evidence to back up their ideas from other sources?

## CROSS-CURRICULAR LINKS

Science: healthy living. The essentials to sustain life – food, water, shelter.

# YEARS
# 5 & 6

## Title
Rivers

## Year
5/6: Autumn Term

## Cycle
A

## Place
A named river, ideally in your locality

## Focus
Rivers and their effects on the landscape

## National Curriculum
1a,b,d
2a,b,c
3a,b,c,d,e,f
5a,e
6
7a,b

| PROGRAMME FOR GEOGRAPHY (TASK 1 THE OVERVIEW) | | | | | | |
|---|---|---|---|---|---|---|
| Y | 1 | 2 | 3 | 4 | 5 | 6 |
| ONGOING STUDY – WEATHER | The locality of the school | The locality of the school | The locality of the school | The locality of the school | Thematic Study | Thematic Study |
| | | | Weather | | | Here, there and everywhere |
| | Thematic Study | Contrasting locality overseas | | Locality in South or Central America, Asia or Africa | Contrasting locality in the UK | |
| | A | B | A | B | A | B |

## Useful resources
Local maps at a range of scales. Reference books (e.g. Wayland series; see page 96). Information and maps for your area (available from the National Rivers Authority and your local water authority).

## Suggested activities

### STARTING POINT

A visit to your nearest stream or river, or another suitable named river. (Clear photographs could be used if a visit is not possible).

### SAFETY

What are the school's policies on planning and organising school trips? Consult local authority guidance on field trips.

### MAKING OBSERVATIONS

Look at the river channel. Is it straight or curved? From a suitable bridge, measure how wide the river is. Which direction is it flowing? Find out by throwing a float into the centre and observing it float away. Drop sticks into different parts of the river and observe whether they float in the same direction. Is it flowing as fast at the sides as at the centre? Use a compass to help determine the direction in which the river flows.

Choose a straight reach of channel, 5 or 10 metres long. To measure the movement of the water, time floats in seconds. Explain the need to undertake a fair investigation by taking the mean of three readings.

Is there any evidence to indicate that the river varies in level? Are there any river deposits? Children should consider where these are in relation to the shape and flow of the river. Are there areas where debris or rubbish collects? What type of debris or rubbish? Where may this have come from?

Is there evidence of the river being used? If so, how and by whom? What evidence is there of other river life? Look at the river banks. Are they steep or gentle? High sided? Have the sides been made artificially higher? Can the children think of reasons for any of their observations?

Look at the ground around the river sides. Is it flat, gently sloping, undulating or hilly? Where is the river in relation to the surrounding ground (at the lowest point)? How is the land on the river banks being used? Are both banks similar or used for different purposes?

## MAKING SKETCHES

Discuss the types of information that it would be useful to sketch to give as full a picture as possible to others of what the river is like. Children could complete a simple land-use map for the river bank area seen.

## IN THE CLASSROOM

Investigate how rivers erode, transport and deposit materials using a large size sand tray or water trough filled with sand and gravel (ballast is ideal). Pebbles can also be used. Raise one end very slightly and use a small watering can to show how the 'river' moves sand downstream. Watch carefully the difference between small and larger pebbles and compare with the sand. What difference does raising the higher end make to the movement and river course?

Use a long piece of plastic guttering (at least 4 metres) with some ballast placed along its length. Raise one end, bending the guttering to leave half of it flat on the ground. Slowly sprinkle water into the raised end. Watch what happens to the sand and gravel as the water reaches the flat part. Which is deposited first? What is carried furthest? Relate each of the findings to the behaviour of the river.

What landscape features result from the action of rivers?

## USING MAPS

Use a suitable map to locate the river visited. You may need to help the children locate this river on a smaller-scale map. It is important to have the opportunity to trace the river back to its source. Follow the river's course to its mouth. Introduce the term tributary: it may be possible to find several tributaries for the main river. What is the name of the river? Which sea or lake does the river flow into? Use information and a map provided by the local water authority to find out the catchment area for your river. (The children may be surprised at the extent.) Relate the extent of the catchment area to a physical map of your part of the UK, taking particular notice of the hilly and lowland areas. Where does the water come from and go to?

## MAKING COMPARISONS WITH OTHER RIVERS

Identify other major rivers within the UK, Europe, and other parts of the world on Maps A, B and C and others. Using atlases, find the source of each of these rivers. Into which sea does each flow? Mark them on a class maps of the UK and the world.

Let the children choose a river from the UK, Europe or other parts of the world to research. Encourage them to focus on the particular landscape features associated with the river that they have chosen. Make a river database using *Junior Pinpoint* on Archimedes.

## CROSS-CURRICULAR LINKS

Science – collect water samples on the field trip. Explain that samples should be collected with care to ensure the environment is not damaged. If possible, carefully collect a small sample of water with part of the river bed or side. (It may be safer for a teacher to do this.)

## Title

**A named issue**

## Year

5/6: Autumn Term

## Cycle

B

| PROGRAMME FOR GEOGRAPHY (TASK 1 THE OVERVIEW) | | | | | | |
|---|---|---|---|---|---|---|
| Y | 1 | 2 | 3 | 4 | 5 | 6 |
| ONGOING STUDY – WEATHER | The locality of the school | The locality of the school | The locality of the school | The locality of the school | Thematic Study | Thematic Study |
| | | | Weather | | | Here, there and everywhere |
| | Thematic Study | Contrasting locality overseas | | Locality in South or Central America, Asia or Africa | Contrasting locality in the UK | |
| | A | B | A | B | A | B |

## Place

A suitable identified locality, perhaps near the school

## Focus

Attitudes to settlement change

## National Curriculum

1a,b,d
2a,b,c
3a,c,d
4
5a,c,d,e
6
9c

## Useful resources

Newspaper articles about a local issue. Local maps.

## Suggested activities

### STARTING POINT

Ideally, a local newspaper report on a land use issue.

### IN THE CLASSROOM

Issues can arise, in both rural and urban environments, where people's expectations and demands in terms or leisure, recreation or the provision of services may be at variance with, for example, local people or landowners – particularly where economic factors are part of the debate. Examples of such land use issues might be the siting of caravan and camping sites, a proposed by-pass or road-widening scheme or the provision of a landfill site.

If possible, use a local issue as the focus for the activities. This can be one being debated at the time or one that has recently been resolved. If this is not possible, then create an imaginary scenario appropriate to your local area. This activity has been written for use with an imaginary scenario. Whatever the focus of your study, it is important to handle it sensitively and encourage open and fair-minded discussion.

Set the scene for the children: a local landowner has cleared some land in a residential area and put it up for sale. Identify the area to be discussed on a local map. Discuss the land use around the area and what road access there is to the land. What ways can the children suggest that the land could be used? Discuss who might be pleased if the land was to be used in the ways they suggest.

Draw up a list of 'interested parties' – imaginary local characters who would have a definite point of view regarding any use of land in the locality. (The list on the next page could be used as it stands, or expanded upon.) Either give each child a 'character' from the list to role play, or split the children into groups, each representing one 'character' and his/her 'supporters'. Give the children time to work out their point of view, develop their characters, and note down their ideas. The groups who wish to develop the land can make a map or plan to show where the features will be sited.

Interested parties (the children may suggest other groups or individuals):

■ landowner (wants a good price for the land);

■ estate agent (needs to fetch a good price for the commission);

■ people who live in the area (want better services/worried about further traffic);

■ parents (want open land/play areas for children);

■ local shopkeepers (competition will force them out of business);

■ supermarket owner (new supermarket to the area);

■ manufacturing industry (bring new jobs to the area);

■ local council (received request from locals for better services);

■ builder (lots of houses for first-time buyers);

■ young couple renting a room (need a place of their own they can afford or need to stay in area near family and job).

At a meeting with representatives of the local council, the children discuss – in character – the situation, each trying to persuade the councillors of the validity of their point of view.

### USEFUL TIP

Unless your children are already adept at simulations you will need to chair the discussions. This is a complex scenario, so do not worry if it cannot be easily resolved.

### CROSS-CURRICULAR LINKS

English.
Drama.

## Title
**Here, there and everywhere**

## Year
5/6: Spring Term

## Cycle
B

## Place
Named places in UK, European Union and the world

## Focus
Placing places

## National Curriculum
1a,b,c,d
2a,b,c
3d,e,f
5b,e
6

| PROGRAMME FOR GEOGRAPHY (TASK 1 THE OVERVIEW) | | | | | | |
|---|---|---|---|---|---|---|
| Y | 1 | 2 | 3 | 4 | 5 | 6 |
| ONGOING STUDY – WEATHER | The locality of the school | The locality of the school | The locality of the school | The locality of the school | Thematic Study | Thematic Study |
| | | | Weather | | | Here, there and everywhere |
| | Thematic Study | Contrasting locality overseas | | Locality in South or Central America, Asia or Africa | Contrasting locality in the UK | |
| | A | B | A | B | A | B |

## Useful resources
Maps of UK, Europe and the world, including Maps A, B and C. Atlases. Pictures, postcards and photographs of named places and features, including those related to topical news items and Maps A, B and C. Travel information. Globes. *Geosafari*. Bus and rail tickets and timetables. Newspapers. Football programmes. Rock samples. Reference books on geography. Oblique aerial photographs.

## Suggested activities

**STARTING POINT**

A collection of interesting artefacts from the UK, European Union and around the world. Encourage the children to add to the collection. (The collection could have a theme to it – e.g. food from around the world.)

**IN THE CLASSROOM**

Can we locate, on appropriate maps, where each of the items has come from? It is important to have a context for introducing the name of a place. Children will remember names of places and their location much more easily if they can see the reason why it has been brought to their notice.

Where children have heard of places, say in the news, then encourage them to take the initiative and use an atlas to find the places. A large map on the wall with 'Places we know' marked, or personal maps of 'Places I know' can be built up over the term. Ideally some information, written or visual, will be linked to each of these places. This may be an item from a newspaper or a postcard, for example. Each case provides an opportunity for reinforcing locational information. Encourage the children to draw on their knowledge and understanding about places and themes from previous work. Encourage children to use the evidence to consider how similar/different from each other the localities are.

A globe will always provide a more accurate image of the world in terms of the spatial relationships between places, so should be used once unfamiliar names of places or features have been identified.

## USING INFORMATION TECHNOLOGY

Children enjoy a 'game' approach to place identification. *Geosafari*, an electronic identification game, can be used with teacher-produced cards, allowing you to select which places you want the children to identify. This is particularly helpful when identifying points of reference on Maps A, B and C.

Extend the activity by focusing on the world's main rivers, weather patterns, uses of land and environmental management. Encourage the children to recognise patterns and to consider reasons for these patterns. CD-ROM encyclopaedia, TV programmes and videos may be used to provide more interest and information about places and themes.

Everyday materials, such as bus and rail tickets and timetables, newspapers, football programmes, rock samples and food can also provide interesting starting points for further geographical work.

## Title
**Contrasting locality in the UK**

## Year
5/6: Summer Term

## Cycle
A

## Place
A named forest or wood

## Focus
Investigating a changing environment

## National Curriculum
1a,b,d
2a,b,c
3a,b,c,d,e,f
4
5a,b,c,d,e
6
10a,b

| | PROGRAMME FOR GEOGRAPHY (TASK 1 THE OVERVIEW) | | | | | |
|---|---|---|---|---|---|---|
| Y | 1 | 2 | 3 | 4 | 5 | 6 |
| ONGOING STUDY – WEATHER | The locality of the school | The locality of the school | The locality of the school | The locality of the school | Thematic Study | Thematic Study |
| | | | Weather | | | Here, there and everywhere |
| | Thematic Study | Contrasting locality overseas | | Locality in South or Central America, Asia or Africa | Contrasting locality in the UK | |
| | A | B | A | B | A | B |

## Useful resources
Outline map of the UK (template 6 on page 106). Forestry Commission pack for schools (see page 90). Information and maps of selected woodland.

## Suggested activities

**STARTING POINT**

A collection of items made of wood (including paper and/or card).

**IN THE CLASSROOM**

Begin the project with a general discussion about what the items are made of, whether they are fashioned by hand or machine, whether they are natural in colour or altered by, for example, staining. Do any items have the country of origin marked? If so, where are the countries on the world map?

Discuss where wood comes from. Where have the children seen woods? Which type of trees are cut down to make wooden items? Who cuts down trees?

Why are woods important? How are they managed and sustained? Does woodland need special protection? Consider environmental threats such as acid rain and pollution, the need for conservation and recreational use of woods. The information from the Forestry Commission will help the children to consider these points.

**MAKE A VISIT**

If possible, visit a local woodland or forest managed by the Forestry Commission. Many have forest trails and some have facilities for school parties. Contact the Education or Recreation Ranger. Teachers' packs or information sheets are often available from the Ranger.

Where is the wood? What route should we take to get there? How accessible is the wood for visitors? If possible, use a survey to find out which areas the visitors come from. How often do they come here? Back in the classroom, draw a signpost map to show where visitors are from and graphs showing reasons for visiting and numbers visiting once, more than once and regularly.

What facilities have been provided for visitors? Where are the facilities located within the wood? Is information provided for visitors on site? If so, where is it? What evidence is there for recent changes in the wood? Make a map of the route taken through the wood, indicating visitor facilities, new planting, tree felling and features of special interest (e.g. a pond, open area).

In the wood, the children should look for evidence of : Who works in the wood? What type of work do they do? Who uses woods? How are they used? What changes would 'improve' the wood?

## USING MAPS

What is it like in the wood? Use your senses to explore the wood. Consider the ground area and tree tops too. What size is the wood? Use a 1:10 000 or 1:25 000 map to compare the area of land covered by woodland with the area covered by your locality. Compare the symbols used for woodland on different size and types of maps, including old maps. Has the size and shape of the wood altered in the last hundred years? What might be the reason for this? Discuss what other changes might have resulted from people 'using' the woodland since the old maps were drawn up, based on their visit.

Extend the activity by using maps to find out where large areas of forests are in the UK. These could be drawn on the outline map. Compare their location with a physical map. Which of the UK countries has the largest areas of forest? Can the children detect any pattern?

## CROSS-CURRICULAR LINKS

Science AT2.

Please note: the term 'wood' has been used to represent either a wood or a forest in these activities.

Good geographical teaching and learning uses the enquiry process to motivate children to find out about the physical and human environment, and their relationship with it. The enquiry process basically means finding out answers to questions. Questions must be formulated to help develop geographical ideas and concepts, and to explore places and themes. Effective questioning enhances the quality of geographical education. It develops attitudes of curiosity and interest, and it ensures an active involvement in the learning process. Relevant and well constructed teacher-led questions in the early stages will serve as a model for children's enquiries. We should encourage children to ask questions and search for answers, based on what they already know and from data sources.

Children should be regularly involved in simple classroom enquiries, perhaps using photographs as a stimulus, as well as those in the school grounds and those involving field work away from the school itself. The whole class may be involved in contributing to a particular enquiry or there may be more than one enquiry being undertaken, relating to a particular place or theme. Enquiries may therefore be short, or take a longer period of time. Initially you will lead the enquiries, though children should be involved in the collection of information and in any discussions about how best to communicate that information from the outset. Gradually, as the children's skills in the enquiry process develop, they will be able to assume greater responsibility for the enquiries.

The following step-by-step approach will help you to plan for a geographical enquiry with your children.

## A step-by-step approach to the enquiry process

1. What are we (the children) going to find out about? (linked to the Geography Programmes of Study for a Place and/or Theme)

2. What question(s) do we need to ask?(specific geographical question)

3. What information will we collect?

4. What does this information tell us?

5. What have we found out? Why is this so? What can we do about it?

6. How can we best show what we have done?

7. Was this helpful/useful/relevant? What should we do differently next time?

# ... developing map work

Children need the opportunity to see and use a range of maps at Key Stages 1 and 2. All maps are selective in their content, showing what the mapmaker wants us to see. It is important, therefore, for schools to build up a collection of a wide variety of maps to support and enhance geographical work in their school, including, for example, pictorial maps, simple teacher-drawn maps, historical maps and computer mapping programs as well as Ordnance Survey (OS) maps. Some of the maps we collect may be free (e.g. tourist-type maps) or available at little cost (e.g. newspaper maps).

The varying styles of maps will help children in their own map-drawing. Different styles are appropriate for different purposes, as shown by the children's maps illustrated. These also show a fairly typical progression in children's drawing of maps, from:

- a simple route map with stylised houses serving as symbols (**Map 1**); to

- a more detailed route map with a wider variety of symbols used (**Map 2**); to

- a village plan showing particular locations using a number-coded key (**Map 3**); to

- a small-scale map showing some indication of more standardised use of symbols (**Map 4**).

**Map 1**

**Map 2**

**Map 3**

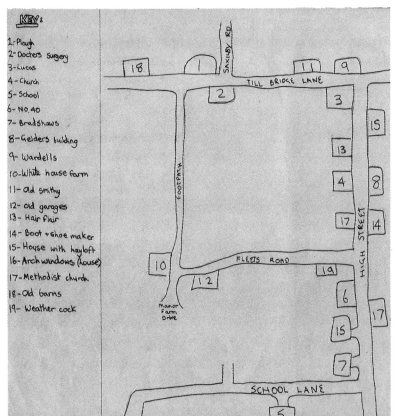

KEY:

1- Plough
2- Doctors surgery
3- Lucas
4- Church
5- School
6- No. 40
7- Bradshaws
8- Gelders building
9- Wardells
10- White house farm
11- Old smithy
12- Old garages
13- Hair flair
14- Boot + shoe maker
15- House with hayloft
16- Arch windows (house)
17- Methodist church
18- Old barns
19- Weather cock

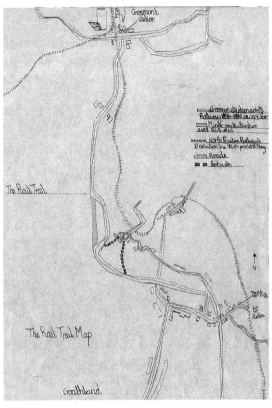

**Map 4**

As a very general rule, the older the child, the smaller the scale of the map. (Large-scale maps show small areas of ground; small-scale maps show large areas of ground.) Certainly younger children find it easier to use maps showing little information, ideally with only the information required for the activity being undertaken. However, often older children will continue to use, for example, a map of the school grounds because it is useful in their continuing studies, while younger children will be interested in maps such as a road atlas which show a journey they are about to make or have recently made.

Ideally, all schools should have the following range of OS maps a\ their locality:

1:1250 (1cm to 12.5m; only available for urban areas)
1:2500 (1cm to 25m; available for rural and urban areas)
1:10 000 (1cm to 100m). (This map would also be particularly useful fo.
                  named contrasting locality study in the UK)
1:25 000 (4cm to 1km; Pathfinder series)
1:50 000 (2cm to 1km; Landranger series).

Large-scale maps of town and city shopping centres are available from Chas. E. Goad (for further details see page 90). In addition to these, schools will need a map of their school grounds. If one is not already available, then it may be useful to have it mapped by a mapping orienteer. The map could then be used for outdoor and adventurous activities in PE as well as geographical studies. Other maps will also be required for your specific place studies. These maps will be in regular use and will need to be laminated to protect and preserve them. OS maps are subject to strict copyright regulations. Some can be copied under certain restrictions and OS can advise on this. Your Education Authority may have a licence to copy maps under certain restrictions, and may also be able to help you with information regarding copyright.

Oblique aerial photographs, such as the one on the cover of this book, are needed for Key Stage 2. Ideally children will initially use those showing their locality in order to build up their skills in using aerial photographs. Wherever possible, aerial photographs should be used alongside appropriate maps of the area shown at a similar scale. Aerial photographs for most of the UK can be obtained from either OS, Photo Air Ltd. or NRSC Air Photo Group (for further details see pages 91-92).

There have been several atlases produced since the introduction of the National Curriculum for Geography which have a style and layout which is attractive to children, and the maps are less cluttered than the more traditional style. Beware of buying large quantities of one type as all atlases have their limitations. The Geographical Association can provide useful information about the most popular atlases for Key Stage 2. Remember that you will need at least one copy of a world atlas in your school which has a great deal of detail, in order to identify smaller places and features which children may hear about and want to locate.

Whenever world maps, or maps of continents are used, globes should be used alongside to give a more accurate image of shape, size and positional relationships between continents and countries of the world.

An overhead projector can be used to create large-sized maps very cheaply by drawing the outline of the map required on an overhead transparent sheet and projecting the image onto a suitable sheet of card on a wall. Simply copy the outline onto the card.

# . . . developing field work

Field work includes all geographical work undertaken outside the classroom, including in and around the school grounds and its immediate locality. It involves studies of both human and physical environments as well as visits to places of geographical interest. Field work should be viewed by all schools as a vital aspect of geography at both Key Stages 1 and 2.

Field work . . .

■ Helps to equip children with a variety of geographical skills.

■ Enables children to become concerned with issues.

■ Extends the world with which the child is familiar.

■ Motivates.

■ Is education for citizenship, which is central to primary education.

■ Is active learning.

■ Is enjoyable.

■ Can be fun.

■ Can be exciting.

■ Helps develop a feeling for the environment.

■ Aids children's personal and social development.

. . . and good practice requires it!

Field work provides a context within which children can work, through first hand experiences. Through field work children should develop a greater understanding of how we use the environment, the way in which we work and live within it and develop a greater sense of responsibility for the environment.

The success of any field trip is dependent upon the degree to which the teacher has planned the work to be undertaken both in and outside the classroom. Careful planning beforehand will help to ensure the success of your field trip.

## Preparing for the field trip

Teachers will need to consider the following before taking any children on a field trip:

■ What are the geographical aims for the field trip?

■ The geographical focus for each aspect of the trip. What are the children to observe, to focus their attention upon, and how will their knowledge and understanding be increased?

■ The school's guidelines or policy on field trips/school visits.

■ Any county guidelines on taking children on field trips/school visits.

■ Any guidelines set down by a professional association.

■ Keeping parents fully informed.

■ Ensuring an appropriate pupil/adult ratio.

■ Preparing any adult helpers that may be responsible for groups of children.

■ Explaining to children expectations for the social aspects of any field trip. (Behaviour and discipline are vital aspects to the success of any field trip and children should be clear about how the school expects them to behave throughout the trip.)

■ The need to conserve the sites visited.

■ The need to create and preserve good relationships with the public.

■ Gathering together suitable equipment and resources.

■ Ensuring that any speakers are fully aware of the age of the children and any relevant experience and knowledge that they may have acquired which relates to their talk. (Some speakers may need guidance on what to say and to focus upon.)

# Work inside the classroom

Decide beforehand what will need to be explained or taught to the children in order to equip them to learn as much as possible from their field trip. Identify the skills the children will need to practice in the classroom so that they are able to observe and record, in a systematic way, information that will prove useful back in the classroom. Take time to explain how equipment that will be taken on the trip will be used, and provide opportunities to practice using the equipment.

Decide the aspects of geography that the field trip will focus upon and the background knowledge the children will require for each aspect. If the children are to listen to experts talking about an aspect of an environment they are to visit, then they may need to be given some background knowledge in advance, in order to understand more easily what the speaker is saying.

Obtain appropriate maps so that children can observe and use them to familiarise themselves with the area to be visited. Use different scale maps to build up the children's understanding, and allow them to make comparisons between their local area and the locality to be visited.

# Work outside the classroom

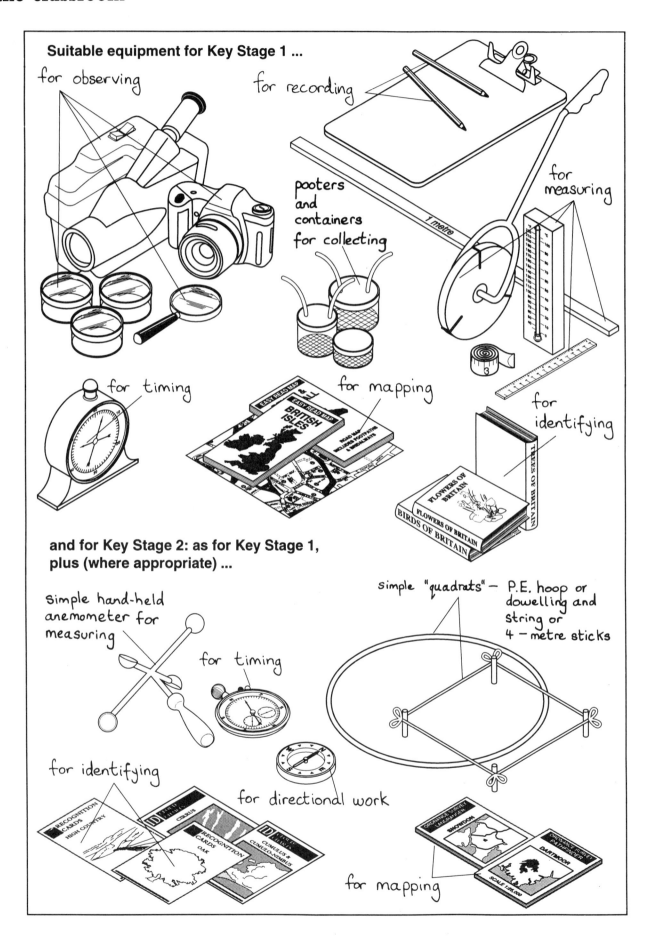

Suitable equipment for Key Stage 1 ...

for observing

for recording

pooters and containers for collecting

for measuring

for timing

for mapping

for identifying

and for Key Stage 2: as for Key Stage 1, plus (where appropriate) ...

simple hand-held anemometer for measuring

simple "quadrats" – P.E. hoop or dowelling and string or 4 – metre sticks

for timing

for identifying

for directional work

for mapping

### USEFUL READING

The Geographical Association provides some useful booklets in their *Field Work in Action* series which give more detailed information on fieldtrips. A number of books dealing specifically with field work are also available. (For further details see Section 7, pages 91 and 97.)

### STUDYING THE LOCALITY OF THE SCHOOL

Particular emphasis has been given to the locality of the school at Key Stage 1, as children need to be able to relate subsequent experiences of places to one which they know well. It is also more accessible for the development of geographical skills, providing direct experience, opportunities for field work and a context for practical activities outside, as well as inside, the classroom. In order to maximise its potential for geographical study, teachers need to be familiar with their local area.

Consider what features your local area has:

1. What landscape and water features does it have?

2. What buildings are there?

3. What evidence is there of its origins?

4. What shops does it have?

5. What evidence of industry is there?

6. What types of transport are in evidence?

7. What services are available?

8. What leisure facilities are there?

9. What local issues are there?

Schools will find it helpful to build up their own resource pack of information for the local area. Some of the information will compliment a History local study unit. Everyone in the school should be involved in this, especially those who live in the area.

Some sources of local information include:

■ List of local people with particular interests or specialisms prepared to talk to the children.

■ Local maps, including historical maps.

■ Photographs (ideally at least A5 size) of particular features and landscape views.

■ Aerial photographs.

■ Service industries information, e.g. gas, electricity, water.

■ Bus timetables and details of other transport services.

■ Articles from the local paper, detailing interesting issues and photographs.

■ Any information provided for visitors to the area.

# . . . using information technology

The following equipment has been found to be the most suitable for use in primary schools. Not every suitable information technology application has been included but those that have explicit links with geography are listed, particularly those which link with the activities set out in Section 5 of this book.

## Using a weather station

Weather is a popular project in primary schools and can be undertaken over any chosen period of time and at any time during the school year. Weather projects offer opportunities for children to take responsibility for observing and recording data, and to observe changes that occur over a period of time.

If preferred a weather project can be referred back to and expanded upon on several occasions during Key Stages 1 and 2. Older children can be given the opportunity to work with more sophisticated equipment and to record different aspects of the weather.

At the upper end of Key Stage 2, children can move from manual recording of the weather to using computers. There are several weather stations available to suit all computers and situations. Schools should ensure that the equipment is not too sophisticated for use by primary children, and that the format in which the data is presented is sufficiently easy for children to understand and interpret.

The weather station kits usually come complete with recording equipment which has to be mounted on a site which is suitable to record the weather and yet not be interfered with by passers by or vandals.

Once your weather station is sited and connected, you may find that the computer cannot be moved without the loss of valuable data. It is therefore worth checking, before purchasing, if the station requires a permanently sited computer and whether the computer can be used for other purposes once connected.

The *Weather Reporter* is one such suitable weather station; and it can be used with most computers in use in primary schools. The recording equipment takes regular readings at 9am each day over a period of 60 days, the reason being that the Meteorological Office takes all its readings at 9am UTC (Unified Time Constant) each day. All meteorological offices around the world use UTC time to ensure parity of readings. Weather recordings for the last sixty consecutive hours are also stored.

Recordings are taken for:

■ wind speed

■ wind direction

■ temperature

■ light

■ rainfall.

From these it calculates maximum and minimum temperatures and wind speeds, day lengths and predominant wind direction.

An extra pack can be purchased which provides the facility to measure humidity and air pressure.

The software provided works on BBC B, A3020, A4000, A5000, Nimbus and IBM PC computers. It will also run on an Archimedes A3000 computer, provided a serial port is fitted.

## Using a monitoring box (*LOGIT LIVE*)

This is a simple data-logging device that can be purchased for use with either a BBC or an Archimedes computer. Once set up in the classroom, it monitors temperature and light, but a wide range of additional sensors can be purchased, including those measuring humidity or oxygen content for pollution sampling. The monitoring equipment can be set up to run for one minute or up to seven days. (If set up and left to run for a couple of days, data can then be displayed on the screen and printouts taken for the children to analyse.)

Set up the *LOGIT LIVE* Box in the classroom to monitor just light and temperature over a 24-hour period. Create a printout of two graphs, one showing light and one showing temperature, and ask the children to explain what each graph is about. Can they explain any sudden changes in the readings? Possible reasons might be the central heating coming on or going off, or sunset and sunrise. Stories can be written entitled 'Our classroom at night'! Readings can also be taken at different times of the year, or in different locations, and the results compared. Separate software called *Junior Insight* will need to be purchased from Longman Logotron if you want to use *LOGIT LIVE*.

## Using weather programs

*My World* is an early years program containing a large number of Key Stage 1 activities. However there is one program on the disc, called *WeatherMap*, which can be used to create a map of Great Britain and Ireland, along with a range of weather symbols. Children can check newspapers each day to obtain weather information, with which to create their own 'weather forecasts'.

The program allows text to be entered on the map so that the children can write down the dates of the weather readings. The maps can then be printed out and kept to look back on and discuss at the end of the term or year. Discussion of the maps could introduce children to the idea that, at different times of the year, the wind comes from certain directions rather than others. Questions could be posed for the children to research from the information contained on their printouts (e.g. Does the cold wind usually blow from the same direction?). If you have access to a colour printer, the children can print out their maps in colour.

*62 Honeypot Lane* is a program in two parts. Part one shows the weather changes that take place at 62 Honeypot Lane throughout the year. Children can move the program on by the hour, day or month and see the changes in the weather and environment. Part two allows the children to actually see inside 62 Honeypot Lane, and move around and explore its rooms.

If children are to gain the most from using this program, then the teacher will need to carefully plan a series of questions for the children to answer as they move through the program:

■ At what time of year are the windows not fully closed at night? Why?

■ Plot the changes on the tree throughout the year. Note the changes, and the dates that they occur.

■ Describe 62 Honeypot Lane at certain times of the year and explain the changes. What reasons can you give for these changes?

■ Having moved around the inside of the house, draw a simple plan for others to follow.

## Using a data-handling program

An example of a weather questionnaire produced on *Junior Pinpoint*.

A simple database can be set up by the teacher or the children on the computer so that children can enter various weather observations each day. An example of a weather questionnaire has been included below to assist teachers in planning their own. Once a number of recordings have been made, the information stored can be printed out in various formats and interpreted by the children.

## OUR WEATHER

Take your readings every school day at 1.00pm.

What is the date?                                   ___ / ___ / ___

What is the temperature?                          ___ C

How much rain has fallen
since yesterday?                                       ___ mm

How much cloud is there?  ☐ No cloud
                                        ☐ less than half of the sky is cloud
                                        ☐ More than half of the sky is cloud
                                        ☐ Total cloud

From which direction is the              ☐ North
wind blowing?              ☐ North-West        ☐ North-East
                    ☐ West                              ☐ East
                    ☐ South-West            ☐ South-East
                                        ☐ South

A database can be used for recording information on all aspects of geography. Good planning by the teacher at the outset can ensure that the data they input challenges children to pose questions and search for answers from the database.

However, the degree to which the information gathered can provide printouts in a format that can be analysed by children is dependant upon the information initially gathered.

Before embarking upon the drawing up of a database, teachers will therefore need to consider carefully the information gathered, the format in which the information can be printed out and the degree to which the printouts can then be analysed.

The easiest way for children to work on databases is not to create them themselves, but to enter information in a pre-prepared database. To create a database for a local weather study, the following questions are suggested:

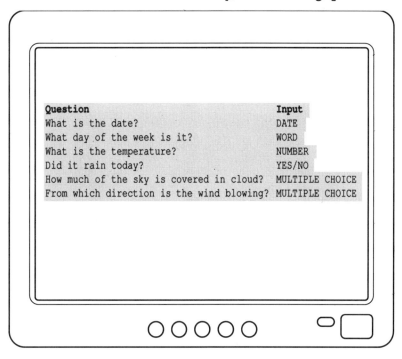

| Question | Input |
|---|---|
| What is the date? | DATE |
| What day of the week is it? | WORD |
| What is the temperature? | NUMBER |
| Did it rain today? | YES/NO |
| How much of the sky is covered in cloud? | MULTIPLE CHOICE |
| From which direction is the wind blowing? | MULTIPLE CHOICE |

## Using databases

Work produced by Year 2 children using *Flare* and *Pendown* to show their views on the destruction of the rainforest.

*DataSweet* is a suite of programs. There is a simple spreadsheet called *DataCalc*, a data collecting program based upon the card file index principle called *Data Card*, and a graphing program called *Data Plot*. This package allows children to create four different kinds of graphs, bar charts, line bar graphs, pie charts and line graphs. These graphs can be saved and placed in *Pendown* – a word-processing package. The spreadsheet is ideal for keeping weather reports.

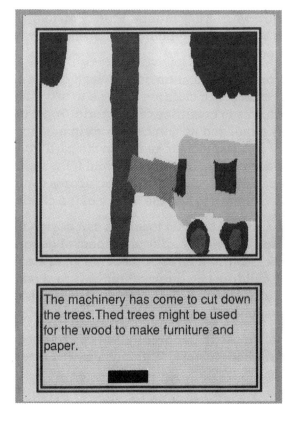

The machinery has come to cut down the trees. Thed trees might be used for the wood to make furniture and paper.

The rain washes away the soil because there are no trees and roots to hold the soil in place.

→

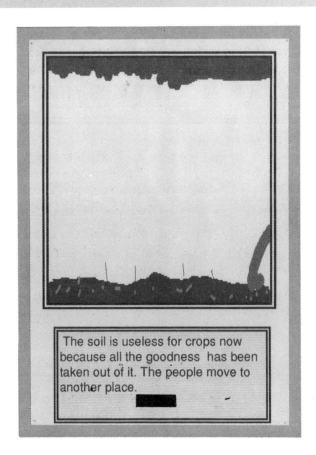

The soil is useless for crops now because all the goodness has been taken out of it. The people move to another place.

*Junior Pinpoint* is a more powerful database, specifically designed for primary children for use in Archimedes on a PC computer. It allows the teacher or children to create a questionnaire from which the database is automatically created. The database can be sorted and searched, and graphs can be plotted. Text, lines and pictures can be added to the graphs so that children have a finished report to illustrate their findings.

## Using Geosafari

This is a mini-computer, which runs independently on four small batteries. It enables children to learn and to reinforce skills and knowledge of different aspects of geography.

*Geosafari* is an electronic identification game which comes complete with a number of coloured cards which children can use to test their knowledge of everything from landforms to countries of the world. Topic cards can be placed on the display panel and its number programmed in. Children are then asked to identify geographical features on the card from those listed on screen. Instant feedback is given, telling the child if their choice is correct or not. Two children can work together and their scores can be recorded. Response times can be altered to suit the ability of the children.

Blank sheets are also provided so that teachers can design their own cards which test the children's knowledge of some aspect of geography recently covered. Cards can be designed on aspects of the local area or on a country being studied. Older children can design their own card games for others to work on. They can also program *Geosafari* themselves, so that it recognises correct responses.

*Geosafari* is not limited to use in geography, but can be used for skills/knowledge in reading, History, Mathematics, and Science.

## Using Roamer

*Roamer* is a floor robot which children can program to follow a specific route. It provides opportunities for children to explore direction and distance. Teachers can create a map of the school locality and children can program *Roamer* to travel in the correct direction to reach its destination. As children become used to programming *Roamer*, other children can provide the directions to follow.

## Mapping programs

*Mapventure* is a useful and popular program for developing mapping skills for use on a BBC computer. Part one focuses on location and shows how maps are created using the idea of riding in a hot air balloon. It draws upon understanding of coordinates and contours for providing locational points.

The teacher can set the program so that it supports work within a particular area. For instance, the coordinates children use can be defined to match their skill level.

In part two children have to find their way to the treasure using views from the four compass directions in each location and a map. Children provide directions around the map using the eight points of the compass.

## Photo CD

It is possible, at a reasonable cost, to take photographs of a local area and have them placed on a Kodak Photo CD by a Kodak franchise dealer. Many primary schools now have CD Rom systems on which these images can be used. Once loaded, children can use the images of their locality as a basis for written work. The images can be put into a word processor or graphics package.

### FURTHER INFORMATION

Further information on most of the information technology applications listed above, together with easy to follow instructions for getting started can be found in the following publication:

*Information Technology*
The Lincolnshire Primary Toolbox
Curriculum and Monitoring Service
Ashleigh Curriculum Centre
Cross O'Cliffe Hill
Lincoln
Lincolnshire LN5 8PN
tel. 01522 544713

See also the list of suppliers on page 94 of Section 7.

# Making assessments and using the level descriptions

Teachers will need to make informal ongoing assessments of children's work in geography, based on learning objectives derived from the programmes of study, in order to inform their day-to-day planning of tasks. Much of this formative assessment may well be held largely in the teacher's head. There is no requirement, at present, to keep evidence of attainment, though teachers may decide to keep a piece of work that typically reflects the standard achieved in geography by most of the class, a group of children or a particular child, to remind them of certain points and to inform the receiving teacher. However, as every teacher will need to make a comment about the achievements of each child in their geography work for the end of each year report, it would seem to be a good idea to make a brief record of this towards the end of the geography studies for that year. The record may make reference to children's knowledge and understanding in relation to local and more distant places, human and physical processes, environmental change and their ability to use geographical skills and information.

The level descriptions are for summative assessment and teachers will need to make judgements relating to these at the end of a Key Stage. To help illustrate how these descriptions are supposed to work we are going to consider one Year 5 child's work. This remains an unreal exercise in that the whole point of summative assessment is that it is based on a teacher's professional judgement. Even so, by discussing the child's work in the context of level descriptions, we can better see what the descriptions actually mean in practice.

## Assessing Sarah's work

Sarah is a Year 5 child of average ability, in a mixed class with thirty Year 5 and 6 children. The material shown on pages 78-79 is a small sample of Sarah's work, which was saved by Sarah's teacher to illustrate how to make an assessment in geography using the level descriptions. The assessment could be used to help the teacher to write about Sarah's progress in the end of term report for her parents.

In making teacher assessments, it is important to bring together all the information known about the child, including:

■ The child's work (which may be a sample of their typical performance).

■ Teacher's observations of the child at work, particularly at the time the work was undertaken.

■ Notes made about the child's work and performance over the Key Stage.

■ Observations and comments from any other teachers who have worked with the child.

In the Autumn Term, the class undertook a local study of North Hykeham, which included elements of both history and geography, with the emphasis on change. Towards the end of the term, the class studied Scott's journey to the Antarctic. The geography focus was on weather in other parts of the world. During the Spring Term, the class studied Kapatalmawa, a locality in Kenya. In the Summer Term, the class made a visit to the contrasting locality of Boggle Hole in Yorkshire.

What judgements can we make about Sarah by looking at her work samples (the written evidence) and by knowing the overall background and context of the work covered? What does this tell us?

After reading the level description for level 2, we can see that Sarah has plenty of evidence here for level 2.

Level 2

Pupils describe physical and human features of places, recognising those features that give places their character. They show an awareness of places beyond their own locality. They express views on attractive and unattractive features of the environment of a locality. Pupils select information from resources provided. They use this information and their own observations to ask and respond to questions about places. They begin to use appropriate vocabulary.

So what about level 3? Does this provide a 'better fit'?

Level 3

Pupils describe and make comparisons between the physical and human features of different localities. They offer explanations for the locations of some of those features. They show an awareness that different places may have both similar and different characteristics. They offer reasons for some of their observations and judgements about places. They use skills and sources of evidence to respond to a range of geographical questions.

Sarah used maps of North Hykenham from 1904 and 1956 to answer the question: 'How has North Hykenham changed during the period 1904 to 1956?' She used mapping skills to draw a map showing the journey from school to Boggle Hole in response to: 'How will we get to Boggle Hole?' Sarah has written about Kaptalamwa, focusing on the differences between Kaptalamwa and where she lives. Her descriptions of Kaptalamwa and Boggle Hole use some geographical terms. The work also shows an awareness of the different characteristics of places: 'In Antarctica the weather is extremely cold, even in Summer'; 'In Kaptalamwa they have very few buildings on [the] landscape'; 'Robin Hood's Bay is along the beach from Boggle Hole'. Sarah used a word processor to record information, and has scanned in her drawing of the Old Farm House.

The level 3 description appears to be a 'good fit', but we need to check level 4, before making a final decision.

Level 4

Pupils show their knowledge, understanding and skills in relation to studies of a range of places and themes, at more than one scale. They begin to describe geographical patterns and appreciate the importance of location in understanding places. They recognise physical and human processes. They begin to show understanding of how these processes can change the features of places, and that these changes affect the lives and activities of people living there. They describe how people can both improve and damage the environment. Pupils draw on their knowledge and understanding to suggest suitable geographical questions for study. They use a range of geographical skills drawn from the Key Stage 2 or Key Stage 3 Programme of Study, and evidence to investigate places and themes. They communicate their findings using appropriate vocabulary.

## BOGGLE HOLE

Boggle hole is near whitby, it has a beach a youth Hostel wore we stayed. Robin Hood's bay is along the beach from Boggle Hole. Boggle Hole got it's name from a cave in the cliffs to your right lived a boggle it is like a drawft and carries a a stick and frightens people. The youth Hostel used to have a beck down beside it and a millers. The millers wife dieded and so did the dog. The millers wife is called the driping lady she hangs around in the games room and you will offer you a loaf of bread

saying hovis on it o if you see the dog it is always howling.

*The Boggle from Boggle Hole*

*The Driping Lady From Boggle Hole*

# Old Farm House

# Middle Street
# Noth Hykeham

## Kaptalamwa

In Kaptalamwa the houses are made out of clay and the roofs are thatched. They have lots of land scape and dusty narrow roads. They have very few buildings on land scape. In Kaptalamwa they do have rain they don't have taps so have to get on a donkey and ride to a stream. The water is used for many important things such as drinking. When chicken go to school they have to take their own books, pencils, pens, rulers and rubbers.

My house is 50 years old it was built in 1939 - 1940. My house was a field before it was built. It has dark orange bricks which have never been changed. It has bright yellow and green moss. And the hooks are rusted and are coming out of the wall. We have a weathered brick which is very old and the mortar is rotting and falling off in between the bricks.

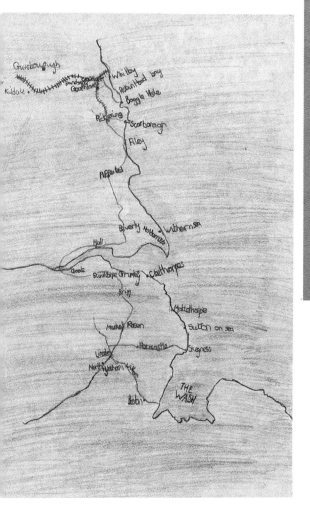

# ANTARCTIC NEWS

ISSUE NO. 5.                                                    PRICE ........P

In Antarctica the weather is extremely cold even in summer. Captain Scott took pemmicann which is dried seal meat. The dogs pulled sledges for Amundson.

The temperature was 0 c and below. The dogs were called huskies and they pulled the sledges. The sledges carried food and tents and drink. They killed the ponies to feed the dogs.

Amundson and his men wore Eskimoes clothes and on their trip to the South Pole and to Antarctica. They ate butter and meat and other food. The dogs pulled the sledges to the South Pole. The weather is freezing.

Penguins are black and white and like very cold places. The temperature is 0 C. There is an ice shelf in Antarctica called the Ross Ice Shelf. Pemnican is seal meat dried.

Well you need food because to live. You need dogs to pull the sledges. You need clothes to keep warm. Well ice is like being cold.

### Antarctica
Anarctica is the coldest continent in the world. Captain Scott sailed to Antarctica, but unfortunately he died of frost bite and scurvy. Long before Captain Cook went to the Antarctic as well, but he turned back. Later some other people went to the Antarctic- as well, to prove that Captain Cook was wrong by saying that Antarctica is not a continent. Sir James Wilkson found the continent but he did not know if it was a coninent. Two years later he died in a plane crash in New Zealand. There was another man called Amundsen. He was very clever indeed. He raced Captain Scott to the South Pole and back, Scott went on a scientific investigation in the Antarctic. Amundsen got back home, Scott did not.

### About Captain Scott
Captain Scott's ship was called the Terra Nova. Captain Scott wanted to get to the South Pole and back again. He was racing against the Norwegians. In 1911 Captain Scott set sail on his long and difficult voyage to the South Pole. On board the Terra Nova was some science equipment. Also on board there were natralists, geologists and meteorolgists. The Terra Nova stopped at seven places before reaching McMuro Sound. Captain Scott took sledges with him. He made the men pull them. Captain Scott thought it was cruel to use dogs to pull the sledges, so instead he kept them as pet dogs. He also brought with him three tractors to test in Antarctica conditions.

Teachers' notes regarding Sarah indicate that she used a globe as reference to write the section for the class *Antarctic News*. She explained to the teacher that the Antarctic is cold because it is a long way from the equator, and also commented in her study on Kaptalamwa that Kaptalamwa was near the equator, and so would be hot. The teacher knows that Sarah has communicated her findings in appropriate language, and that she has studied a range of places at more than one scale. This indicates that she has begun to achieve some aspects of the level 4 description.

However, taken overall, and based on Sarah's typical performance, using written evidence, unwritten professional knowledge of the child and formal records, level 3 appears to provide the 'best fit' for Sarah.

The teacher also notes that next year, when involved in geographical studies, Sarah will need:

■ More opportunities to make comparisons between locations.

■ Opportunities to begin to recognise geographical patterns.

■ To focus on how people affect the environment.

■ To be encouraged to suggest suitable geographical questions for study.

Sarah's teacher recognises that she will be able to use some of this written evidence again to provide judgements in History, English Writing and Information Technology.

# Reference

## Glossary

The following is a list of geographical terms used in this text and in the National Curriculum for geography:

**British Isles**
The group of islands off the north-west coast of Europe. The islands are Great Britain, Ireland, Orkneys, Shetland Islands, Isle of Man, Isle of Wight, Scilly Isles and Channel Islands.

**Catchment area**
An area of land upon which water falls and collects on or under the ground to contribute to a particular river system.

**Channel**
The ground immediately around where the river or stream flows, including the river banks.

**Clinometer**
An instrument used to find the angle of elevation of a slope or to calculate the height of objects, such as a building or tree.

**Contrasting locality**
A small area with distinctive features – similar in size to the locality of the school. The contrast may be in terms of physical features, human features or both.

**Co-ordinates**
A numbers or letters system used to identify points or a square on a grid.

**Deposition**
The laying down of transported material, which includes sharp scree, rounded (water worn) stones, sand and mud.

**Enquiry**
To seek particular information by asking questions which lead children to investigate an issue or idea. An enquiry can use primary and/or secondary sources.

**Erosion**
The action of material carried by rivers, ice, sea or wind to change the shape of the land.

**European Union**
The countries of The United Kingdom, Eire, France, Germany, Italy, The Netherlands, Denmark, Belgium, Luxembourg, Spain, Portugal, Greece, Austria, Finland and Sweden.

**Features**
Distinguishing parts, including built features such as roads and buildings, and physical features such as rivers and hills.

**Field work**
Work outside the classroom, including work in the school grounds.

**Forest enterprise**
One of the three parts of the Forestry Commission. It is responsible for the multi-purpose management of the Forestry Commission's own forests and woodlands throughout Great Britain. Its aims include marketing timber and forest products, increasing opportunities for public recreation, and increasing the attractiveness and conservation value of its forests and woodlands.

**Four-figure grid reference (British National Grid)**
Identification of a square on a grid. Given by reading the two eastings numbers (written along bottom or top edge of the map) followed by the two northings numbers (written along the side of the map). For local referencing, grid letters identifying 100 000 metre square are usually omitted.

**Geographical context**
The way in which places and themes in geography relate to other aspects of geography. Development of the geographical context for places and themes should be gradually increased over Key Stages 1 and 2, including relationships between features both inside and beyond the UK.

**Geographical investigation**
The process of carefully examining a place or aspect of geography.

**Great Britain**
England, Scotland and Wales.

**Landmark**
A prominent feature in the environment, e.g. river, village, wood.

**Locality of the school at Key Stage 1**
The school buildings and grounds and the surrounding area within easy access.

**Locality of the school at Key Stage 2**
An area larger than the immediate vicinity of the school. It will usually contain the homes of the majority of children in the school.

**Map**
A representation of places showing landmarks, features and points of interest. Formal maps show the view from above and follow an agreed set of rules. A map is a cartographer's interpretation of the landscape.

**Mouth**
The point where the river reaches the sea. The river may spread out into several channels, creating a delta.

**National scale**
Covering a widespread area encompassing a group of people or peoples linked, for example, by culture, race or language.

**OS**
Ordnance Survey

**Oblique aerial photographs**
Photographs taken from the air at an acute angle to the ground. Younger children find it easier to identify features on oblique photographs than on vertical photographs.

**Pictorial map**
An informal map with no set rules.

**Plan**
A large scale map or diagram usually used to show a small area such as a room, a building or the layout of a garden.

**Signpost mapping**
Representation of positions in relation to a fixed point. Useful for reinforcing directional work. Signpost mapping may be undertaken in the classroom, school grounds and the local area as well as other places, and using various scale maps.

**Source**
The start of a river or stream (often a trickle of water; its exact position may vary depending on the time of year).

**Transport**
The movement of eroded material. Generally, finer material is transported a greater distance.

**Tributary**
A smaller river or stream joining the main river.

**United Kingdom**
England, Scotland, Wales and Northern Ireland.

**Viewpoint**
The position from where something is seen. Different viewpoints may include near and far, different compass directions, high (e.g. from a hill) as well as ground level.

# Useful addresses

**Action Aid**
3 Church Street
FROME
Somerset
BA11 1PW
Tel: 01373 461623

Action Aid produce several excellent 'locality packs' for Key Stage 2,
including ones from India, Pakistan, Kenya and Peru. They also offer a range
of other educational materials, including stories from around the world.

**British Orienteering Federation**
'Riversdale'
Dale Road North
Darley Dale
MATLOCK
Derbyshire
DE4 2HX
Tel: 01629 734042

The Federation can supply information on all aspects of orienteering,
including details of permanent courses in your area and advice on school
grounds mapping.

**Chas. E. Goad**
8-12 Salisbury Square
OLD HATFIELD
Hertfordshire
AJ9 5BJ
Tel: 01707 271171

Goad plans are plans of shopping centres and high streets. Copyright for
these can be purchased cheaply, and plans from a few years ago can be
used to identify change. A Primary Education Pack, to help teachers using
the plans, is available along with a full range of Ordnance Survey products.

**Commonwealth Institute**
Kensington High Street
LONDON
W8 6NQ
Tel: 0181 603 4535

The Commonwealth Institute can supply information and loan packs relating
to most Commonwealth countries.

**Forestry Commission**
231 Corstorphine Road
EDINBURGH
EH12 7AT
Tel: 0131 334 0303

The Forestry Commission produces a range of useful information packs
(with science links) and *Forest Life* magazine. The Commission can also
supply you with a list of contacts, facilities and available resources for the
forty-five different districts.

### Geographical Association
343 Fulwood Road
SHEFFIELD
S10 3BP
Tel: 0114 2670666

A wide range of excellent, informative publications are available, including the resource packs: *Kaptalamwa, A Village in Kenya* and *Castries, St. Lucia* (in the Caribbean). Conferences (entrance free) are held each year, with primary lectures, fieldtrips and workshops. Annual subscription to the Association includes the quarterly, thirty-six page *Primary Geographer* and other publications at a reduced rate.

### Geosupplies Ltd
16 Station Road
Chapeltown
SHEFFIELD
S30 4XH

Geosupplies supply packs of rocks and fossils.

### Meteorological Office
Marketing Services
Sutton Building
London Road
BRACKNELL
Berkshire
RG12 2SZ
Tel: 01344 854818

The Meteorological Office can offer specific help on weather studies, including supplying weather forecast logging maps.

### NCET
Sir William Lyons Road
University of Warwick Science Park
COVENTRY
CV4 7EQ
Tel: 01203 416994

NCET publications deal with the integration of information technology into geographical work. Data handling packages and other useful software are also available.

### NRSC Air Photo Group
National Remote Sensing Centre
92-94 Church Road
MITCHAM
Surrey
CR4 3TD
Tel: 0181 685 9393

Can supply aerial photographs and a *Discovering Aerial Photographs* Resource Pack for schools. Their education officer can arrange INSET sessions for primary teachers.

**Ordnance Survey**
Romsey Road
SOUTHAMPTON
SO9 4DH

The Education Team Tel: 01703 792795
Customer Information Helpline Tel: 01703 792452

Ordnance Survey produce a huge range of maps at many scales, aerial photographs, CD-ROMs (York is particularly useful) and map skills books. *Mapping News* – available free – keeps schools up-to-date.

**Oxfam**
274 Banbury Road
OXFORD
OX2 7GZ
Tel: 01865 56777

Oxfam produce a range of publications relating to developing countries.

**Photoair Ltd**
191 Main Street
Yaxley
PETERBOROUGH
PE7 3LD
Tel: 01733 241850

Photoair's aerial photographs include county packs (for most counties) and theme packs: cities, transport, physical, rivers & coasts, industry, leisure and villages.

**Silva UK Ltd**
Unit 10
Sky Business Park
Eversley Way
EGHAM
Surrey
TW20 8RF
Tel: 01784 471721

Suppliers of good quality compasses – model 7DNS – which are ideal for children's use. They will also be able to supply copies of *Developing Navigational Skills* – a very useful book for compass work.

# Field work equipment

**Globes**
Can be purchased from many stationers and through educational catalogues. A globe catalogue can be ordered from:

Stanfords
12-14 Long Acre
LONDON
WC2E 9LP.
Tel: 0171 836 1321.

Inflatable globes can be obtained from:

Cambridge Publishing Services
PO Box 62
CAMBRIDGE
CB3 9NA

and Chas. E. Goad (see page 90).

**Playmats**
Wipe-clean maps with land outlines in green and blue for UK, Europe and World are available from NES Arnold.

**Weather measuring equipment**
Thermometers, rain gauges, anemometers (wind speed) and weather vanes are available from:

Invicta Plastics
Oadby
LEICESTER
Leicestershire

TTS Ltd.
Unit 4
Holmewood Fields Business Park
Park Road
Holmewood
CHESTERFIELD

**Simple thermometers** can be ordered from Spectrum Educational Supplies.

# Information technology

*DataSweet*
Kudlian Soft
Hampshire Microtechnology Centre
Connaught Lane
PORTSMOUTH
PO6 4SJ

*Flare*
Silica Software Systems
Mallards
Lower Hardres
CANTERBURY
Kent
CT4 5NU

*Geosafari*
NES Arnold (other packs of cards for various curriculum areas are available).

*62 Honeypot Lane*
Resource
Resource Centre
51 High Street
Kegworth
DERBY
DE74 2DA
Tel: 01509 672222

*Junior Pinpoint* and *Pendown*
(ISBN 0582 100992)
Longman Logotron
124 Cambridge Science Park
Milton Road
CAMBRIDGE
CB4 4ZS
Tel: 01223 425558

*MapVenture*
Sherston Software Ltd
Swan Barton
MALMESBURY
Wiltshire
SN16 0LH
Tel: 01666 840433

*My World*
NW Semerc
1 Broadbent Road
Watersheddings
OLDHAM
OL1 4LB
Tel: 0161 627 4469

*LOGIT LIVE*
Research Machines
New Mill House
183 Milton Park
ABINGDON
Oxfordshire
OX14 4SE
Tel: 01235 826000

Additional sensors are available for *LOGIT LIVE* from
Griffin & George
Bishop's Meadow
LOUGHBOROUGH
Leicestershire
LE11 0RG
Tel: 01509 233344


*Roamer*
Valiant Technology
370 Old York Road
LONDON
SW18 1SP
Tel: 0181 874 8747

*Weather Reporter*
Advisory Unit for Microtechnology in Education
Endymion Road
HATFIELD
Hertfordshire
Tel: 01707 265443

*Automatic Weather Stations*
Producers of a range of weather stations
MJP – Geopacks
PO Box 23
St Just
PENZANCE
Cornwall
TR19 7JS

# Books for children

*In the Forest*. D. Butler. Simon & Schuster. ISBN 0-7500 0286-7.

*Focus on Kenya*. Fleur Ng'weno. Hamish Hamilton Ltd. 1990.
ISBN 0-241-12797-1.

Inflatable Globe for Infants: Collins Longman. ISBN 0-00-360333-4.

*Postman Pat Goes to Town*. John Cunliffe. Scholastic. 1993.
ISBN 0-590-54139-0

*Rivers. The Young Geographer Investigates* Series. Oxford University Press.
1986. ISBN 0-19-917073-8.

*Rivers and Lakes. Our World* Series. Wayland. 1986. ISBN 0-7502-0674-8.

*Rivers, Lakes and Wetlands*. S. McMillan. BBC Books 1992.
ISBN 0-563-36167-0.

*Salaama in Kenya*. Michael Griffin. A & C Black (Publishers) Ltd. 1987.
ISBN 0-7136-2852-9.

*The House on the Hill*. Philippe Dupasquier. Anderson Press.
ISBN 0-86264-167-5.

*Time and Place* Photopack. S. & P. Harrison. Simon & Schuster.
ISBN 0-7501-0085-0

*Water*. J Baines. Wayland. 1991. ISBN 0-7502-0963-1.

*Weather and Climate*. B Taylor. Kingfisher. ISBN 0-86272-942-4.


**Atlases for Key Stage 1:**

Atlas 1. Simon Catling. Collins Longman. 1994. ISBN 0-00-360334-2.

Infant Atlas. Patrick Wiegand. Oxford University Press. 1994.
ISBN 0-19-831687-9.


**Atlases for Key Stage 2:**

A wide variety is available for schools to select from. Publishers include
Oxford, Nelson, Collins Longman, Heinemann Philip, Hamlyn (Giant
Atlases), Folens and Usborne.

# Books for teachers

*A Guide to Field Work in the Primary School*. S. Wass. Hodder & Stoughton. ISBN 0-340-51288-1.

*Discovering Distant Places*. Maureen Weldon. The Geographical Association. 1994. ISBN 0-94-851265-2.

*Field Work in Action: Planning Field Work*. Stuart May, Paula Richardson and Val Banks. The Geographical Association. ISBN 0-94-851261-X.

*Field Work in Action 2: An Enquiry Approach*. Stuart May and Julia Cook. The Geographical Association. ISBN 0-94-851264-4.

*Field Work in Action 3: Managing Out-of-Classroom Activities*. Tony Thomas and Stuart May. The Geographical Association. 1995. ISBN 0-94-851293-8.

*Kaptalamwa, A Village in Kenya*. Maureen Weldon. The Geographical Association. 1994. ISBN 0-94-851266-0.

*Placing Places*. Simon Catling. The Geographical Association.

*Riverwork. An educational resource pack about the water environment*. National Rivers Authority. 1993. ISBN 1-873160-40-2.

*The Outdoor Classroom. Educational Use, Landscape Design & Management of School Grounds*. HMSO. 1990. ISBN 0-11-270730-0.

*Water in the Environment*. Rachel Bowles. The Geographical Association. 1992. ISBN 0-94-851243-1.

# Guides to the curriculum

*Aspects of Primary Education: The teaching and learning of History and Geography.* Department of Education. HMSO. 1989.

*Geography in a Nutshell.* Wendy Morgan.
The Geographical Association. 1993.

*Geography in the National Curriculum.* Department of Education.
HMSO. 1994.

*Geographical Work in Primary and Middle Schools.* David Mills, ed.
The Geographical Association. 1988.

*The Really Practical Guide to Primary Geography.* M. Foley and J. Janikoun.
Stanley Thornes Ltd. 1992.

# PHOTOCOPIABLE TEMPLATES

1 Large-scale grid overlay

2 Medium-scale grid overlay

3 Small-scale grid overlay

4 Weather data sheet: KS1

5 Weather data sheet: KS2

6 Map of the United Kingdom

|  | 1 | 2 | 3 | 4 |
|---|---|---|---|---|
| A |  |  |  |  |
| B |  |  |  |  |
| C |  |  |  |  |
| D |  |  |  |  |
| E |  |  |  |  |
| F |  |  |  |  |

|  | 1 | 2 | 3 |  |
|---|---|---|---|---|

| | A | B | C | D | E | F | G | H | I | J | K | L |
|---|---|---|---|---|---|---|---|---|---|---|---|---|
| 8 | | | | | | | | | | | | |
| 7 | | | | | | | | | | | | |
| 6 | | | | | | | | | | | | |
| 5 | | | | | | | | | | | | |
| 4 | | | | | | | | | | | | |
| 3 | | | | | | | | | | | | |
| 2 | | | | | | | | | | | | |
| 1 | | | | | | | | | | | | |

|  | 1A | 2 | 3 | 4 | 5 | 6 | 7 | 8 | 9 | 10 | 11 | 12 | 13 | 14 | 15 | 16 |
|---|---|---|---|---|---|---|---|---|---|---|---|---|---|---|---|---|
| B |  |  |  |  |  |  |  |  |  |  |  |  |  |  |  |  |
| C |  |  |  |  |  |  |  |  |  |  |  |  |  |  |  |  |
| D |  |  |  |  |  |  |  |  |  |  |  |  |  |  |  |  |
| E |  |  |  |  |  |  |  |  |  |  |  |  |  |  |  |  |
| F |  |  |  |  |  |  |  |  |  |  |  |  |  |  |  |  |
| G |  |  |  |  |  |  |  |  |  |  |  |  |  |  |  |  |
| H |  |  |  |  |  |  |  |  |  |  |  |  |  |  |  |  |
| I |  |  |  |  |  |  |  |  |  |  |  |  |  |  |  |  |
| J |  |  |  |  |  |  |  |  |  |  |  |  |  |  |  |  |
| K |  |  |  |  |  |  |  |  |  |  |  |  |  |  |  |  |
| L |  |  |  |  |  |  |  |  |  |  |  |  |  |  |  |  |
| M |  |  |  |  |  |  |  |  |  |  |  |  |  |  |  |  |
| N |  |  |  |  |  |  |  |  |  |  |  |  |  |  |  |  |
| O |  |  |  |  |  |  |  |  |  |  |  |  |  |  |  |  |
| P |  |  |  |  |  |  |  |  |  |  |  |  |  |  |  |  |
| Q |  |  |  |  |  |  |  |  |  |  |  |  |  |  |  |  |
| R |  |  |  |  |  |  |  |  |  |  |  |  |  |  |  |  |
| S |  |  |  |  |  |  |  |  |  |  |  |  |  |  |  |  |
| T |  |  |  |  |  |  |  |  |  |  |  |  |  |  |  |  |
| U |  |  |  |  |  |  |  |  |  |  |  |  |  |  |  |  |
| V |  |  |  |  |  |  |  |  |  |  |  |  |  |  |  |  |
| W |  |  |  |  |  |  |  |  |  |  |  |  |  |  |  |  |
| X |  |  |  |  |  |  |  |  |  |  |  |  |  |  |  |  |

*KS1 Collecting Weather data*

# What's the Weather like?

| My name is | Today it is | My weather symbol |
|---|---|---|
| **Monday** | | |
| **Tuesday** | | |
| **Wednesday** | | |
| **Thursday** | | |
| **Friday** | | |

*KS2 Collecting Weather data*

# Weather

| Location: | | | | | | |
|---|---|---|---|---|---|---|
| Date | temperature | rainfall | wind direction | wind strength | cloud cover | |
| | | | | | | |
| | | | | | | |
| | | | | | | |
| | | | | | | |
| | | | | | | |
| | | | | | | |
| | | | | | | |
| | | | | | | |
| | | | | | | |
| | | | | | | |
| | | | | | | |